"Exceptionally well written, organized and presented, *This Time It's NOT Personal: Why Science Says Get Over Yourself* is an inherently fascinating read that is as informed and informative as it is thoughtful and thought-provoking. Especially recommended for non-specialist general readers with an interest in science in particular, and how human beings came about and our relative place in the universe, *This Time It's NOT Personal: Why Science Says Get Over Yourself* is very highly recommended and certain to be an enduringly popular addition to school and community library collections."

—*Midwest Book Review*

"A funny, irreverent tour through the salient points of scientific knowledge ... in the best tradition of popular-science accounts from writers like Carl Sagan and Timothy Ferris, taking complex subjects, breaking them down into more basic elements, and presenting them in fast-paced, engaging prose ... the whole vital enterprise of modern science is given a very lively presentation ... a must-read for the scientifically curious."

—*Kirkus Reviews*

Five out of five stars.
—*Readers' Favorite*

# THIS TIME IT'S NOT PERSONAL

## Why Science Says Get Over Yourself

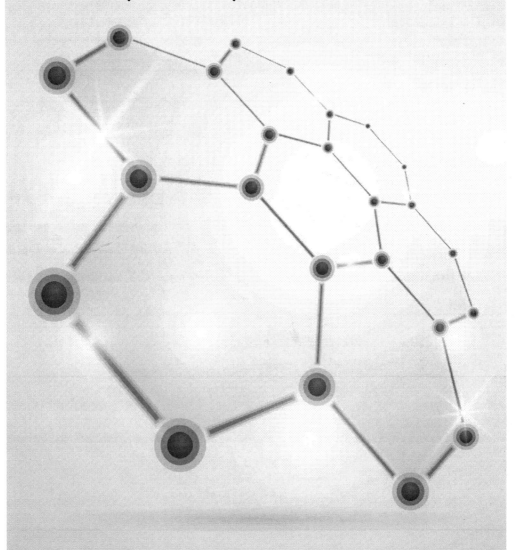

*Sam Hicken*

# ABOUT THE AUTHOR

Sam Hicken, Ph.D. has received two national awards for research that relates to the psychology of computer use. His 1991 doctoral dissertation was designated best in its field. He's been a professor, Director of Informatics for a cutting-edge biotech company, and a consultant to the Infectious Diseases Division of the New Mexico State Health Department. His work spans computer science, cognitive psychology, and molecular biology, plus he's authored dozens of computer databases and two pieces of commercial mathematics software. (*EasyQuant* was one of the first simple-to-use statistics programs for personal computers.) He currently writes scientific and other fact-based humor.

*www.samhicken.com*

ISBN-13: 978-1502892591
ISBN-10: 1502892596

Printed by CreateSpace, An Amazon.com Company
Also available on Kindle and online stores

# CONTENTS

# PREFACE

I've had a problem with reality. There's always been something about existence and how it was portrayed to me that seemed off. As a kid I liked to figure out how gadgets worked, even if my methodology guaranteed they'd never work again, which is one reason my sixth grade teacher, Mr. Spatola, urged me into science (and via his abundant wit, into clownery).

A youthful scientific bent evolved into a career in computers, biotechnology, and education, where for more than twenty years I hewed away professionally and fancifully at the existential conundrum. My fool's errand was to fit together how existence operates, hoping that my methodology wouldn't guarantee its demise. Two and a half years ago I read an article entitled "Why Things Change" by chemist Peter Atkins, and the final piece of my private puzzle fell into place. A blueprint of the objective reality that underlies subjective experience etched itself in my mind, and I wrote this book.

The account gallivants from the Big Bang to the present—not just to our current era—to the here and now. It's an empirical tale, seldom religious, occasionally philosophical, sometimes personal, and hopefully funny. In any case, it's my story and I'm sticking to it.

My goal, by way of science and humor, is to pitch self-*devaluation*. Scientific research has certified the rightful grounds, and levity, I believe, is a sensible response to them. I felt, as soon as the thesis gelled in my mind, that we overvalue self to our detriment, and having now laid out the eclectic evidence I know the thesis to be true. A twenty-five year quest has come to a persuasive conclusion, one that I now must live with.

When Oxford scholar Derek Parfit ceased to fixate on a separate self, he wrote, "My life seemed like a glass tunnel, through which I was moving faster every year, and at the end of which there was darkness. When I changed my view, the walls of my glass tunnel disappeared. I now live in the open air. There is still a difference between my life and the lives of other people. But the difference is less. Other people are closer. I am less concerned about my own life, and more concerned about the lives of others."[1]

Ditto. I don't get as psyched up about my so-called self as I used to. I don't root as hard for the home team. Life feels more equable, but flatter I confess, now that I've relinquished my throne at the epicenter of existence. I couldn't backtrack if I wanted to, because for a scientist the bitterest truth is better than the sweetest lie.

For a humorist, though, truth roosts in the brambles of human paradox, as when humanity achieves planetary ascendance then trashes the planet, or when sailors head off to war playing frisbee on the deck of a battleship. Luckily, when I'm crying on the inside I can (almost always) laugh on the outside. For example, I was tickled by this headline from Treehugger.com (italics theirs): "The Pacific Garbage Patch May Not Be *Twice* The Size of Texas."[2]

A sense of humor helps indeed, as have many of my accomplished friends and colleagues. I'm especially thankful to longtime confidant Bob David, a lawyer, and childhood classmate Drew Tiene, a professor, who perused drafts of the book and chipped in constructive suggestions. My former colleague Carla Westen did the same, enouncing a female perspective that, frankly, I needed to hear.

# INTRODUCTION

*"A human being is part of a whole, called by us the Universe, a part limited in time and space. He experiences himself, his thoughts and feelings, as something separated from the rest, a kind of optical illusion of his consciousness. This delusion is a kind of prison for us, restricting us to our personal desires and to affection for a few persons nearest us."*
*- Albert Einstein*

---

"What is a time machine?"

"A device to visit the past and the future."

"If the past and future exist, why can't we just drive?"

"We can't get there from here."

"Where are we?"

"The present moment."

"If we can't escape the present, what's the use of past and future?"

"Without the past we'd have no identity, and without the future there'd be nothing to protect."

"Would that be bad?"

"Excellent question."

"Is there really such thing as a time machine?"

"There are two:  Doc Brown's DeLorean and the mind."

"Can we test one out?"

"Let's test the first now and the latter later."

Set the DeLorean for 1927.  The Babe, who they say was known for it, would surely hoist a glass with us after swatting his 60th home run, and that same year we could relish a hearty helping of American pride as *TIME* magazine honored its first "Man of the Year," Charles Lindbergh. Each year, *TIME* salutes "a person, couple, group, idea, place, or machine that, for better or for worse, has done the most to influence the events of the year."[3]  *TIME* boldly shattered the gender barrier as early as 1999, introducing its first "Person of the Year" (Jeff Bezos, founder of Amazon).

Pop Quiz:  Who was *TIME* magazine's 2006 Person of the Year? Hint #1:  This individual, near and dear, is sandwiched between *TIME*'s 2005 designees, "The Good Samaritans" (represented by Bono, Bill Gates, and Melinda Gates), and 2007 awardee Vladimir Putin.  Not yet?  Hint #2:  Get up, walk to the bathroom, or bedroom, or likely any room in the house, and gaze in the mirror at *TIME*'s 2006 Person of the Year. YOU! (Not you personally, but each of us individually.)

Award-winning cultural documentarian Ken Burns has observed on several occasions that since 9/11 what the government expects of us, as patriots, is to shop. Another image Burns has brought into focus is a culture of self. "Everything in our media culture is narcissistic, it's all about you. We don't actually experience real life."[4] America's Finest News Source, *The Onion*, spotlighted this story: "Internet Outage Forces Public Into Street To Voice Their Inane Opinions."

Like it or not, if we could objectively view ourselves in the mirror we'd realize there's less there than meets the eye, which is what this book is all about—observing our "selves" scientifically (and rolling our eyes). But even as we come to intellectually accept our relative insubstantiality, we'll still cling to ourselves instinctively, emotionally, subjectively, and protectively, as autonomous VIPs. That's just how nature built us. Even Superman and Superwoman can find themselves holed up in a fortress of solitude.

You might be thinking, "I may not be super, but I'm no me-firster. I happen to care about people." Then let me ask you a question. Hungry children, violence worldwide, planetary pollution, that homeless person at the corner, your friend with cancer—is any of that going to stop you from enjoying your next meal? If you think it's just commonsense to look out for number one, please join me as science demonstrates how that can lead to a minefield of number two.

Selfness, which both scholarship and simple observation reveal is business as usual, spawns documentable, disquieting consequences. One example is the "fundamental attribution error," an unconscious bias that's been repeatedly affirmed in psychological research. Specifically, studies show that we tend to attribute other people's life-standing and behavior to their personal characteristics, whereas we allow ourselves situational justification, particularly for our failings. Overall that simply can't be true, and holding others unduly accountable is surely not the most constructive interpersonal perspective. Moreover, as individual selves aggregate into cliques, factions, and societies, analogous group-based misattributions and their repercussions arise.

We'll review sheaves of evidence in the case of Science versus the Dominion of Self, but to summarize, up front, the Court's key finding: Personal identity as conventionally experienced, though an enduring sentimental favorite, is neither supported by objectivity nor principally beneficial.

When I look in the mirror I fret every blemish, but nobody else seems to notice. True story—I've chopped my hair and shaved off my beard then been asked, "Is that a new shirt?" When psychologist Daniel Simons and his colleague Daniel Levins performed a series of studies in change blindness, a form of inattention, they found that we rarely regard others with the scrutiny we devote to ourselves. In a typical study, an experimenter with a map approaches a passerby to ask directions. After the passerby has been giving directions for about fifteen seconds, two more experimenters, carrying a door, walk between the passerby and the initial experimenter. As they pass, the original experimenter switches places with the experimenter carrying the back end of the door, at which point the substitute direction-seeker carries on the conversation with the passerby. Simons and Levins discovered that in just over 50% of cases the passerby didn't notice he or she was talking to a different person, and that passersby took even less notice of the swap if the experimenters were dressed like construction workers.[5] The takeaway is that unless someone is sexually interested in (or jealous of) you, they're not mapping your blemishes.

This book is a scientific excursion from the origin to the destiny of "moi," the inner Miss Piggy, the singular node of consciousness that peers out from the inmost sanctum with an eye to me and mine. It's about self. Moi.

Our leisurely expedition will meander around and crisscross the scientific turf of physics, cosmology, biology, psychology, mathematics, informatics, and neuroscience. En route, we'll regularly pull over to peek in the window of exotic roadside attractions such as dark energy, the Monty Hall problem, free will, caveman logic, and an item that's now a hotbed of scientific discourse, Neanderthal sex.

With the exception of computers and information systems, I'm not an authority in any of the above fields, hence I've relied on outside experts, some famous, some in print, some on the Internet, and many via my beloved Kindle. The science laid out is neither contorted nor contrived; it was simply culled, cross-referenced, and organized based on how it informs notions of self.

Because science constantly evolves, the evidence on exhibit is as up-to-the-minute as possible. Many of the facts and assertions have been conclusively validated, and in less definitive cases current scientific consensus is reported. I've attempted to point out matters of scientific contention, and areas in which informed speculation is the best we can do, for now.

There are flights of absurdist fancy interleaved with the science, and I've tried to make clear which segments are scholarship and which are caprice. For clarity, lengthier diversions appear within a rectangular frame, as shown.

---

*One Thing Politicians Understand About Factual Information*

If you torture a statistic long enough, you can get it to confess to anything.

---

Some of the science is fantastic to the point of implausibility, so if on occasion I add words like, "I'm not making this up," that section is 100% authentic. When I write, for instance, that according to David J. Linden, a neuroscientist at John Hopkins Medical School, "Humans are truly the all-time twisted sex deviants of the mammalian world," I'm quoting from his book. In short, if an assertion, quote, or passage is not unmistakable whimsy, it's fact.

While the arts testify to inner reality in a profound and satisfying sense, they offer little objective counsel as to the origin, constitution, life cycle, fate, or ultimate significance of personal selfhood. For empiricists and others attentive to the angles from which the spotlight of evidence might illumine the inner, it's sci-way or the highway. Sirens of wishful thinking and narcissism turn out to be just that. Magic, liturgy, and new age fancies are eschewed here. The book's bottom line premise is that self arises through a natural process that can be traced from its origin, through its existence, to its destiny and import.

A greatly compressed version of the story might go as follows. The cosmos spewed out raw materials that led, on at least one planet, to life. Natural selection favored creatures that could exert control over their environment, resulting in increasingly complex nervous systems. Eventually, the advantages and ramifications of escalating brain capacity resulted in not just simple consciousness, but self-awareness.

At least till now, self-consciousness has dished out good news and bad news, a mixed bag of goodies and gotchas. It caresses us tenderly then slaps us across the face. We tango with then arm wrestle ourselves, all the while broadcasting the drama to anyone willing to leave the dial tuned to radio KMOI. Why do we do it? Could it be different? Who are the sages and scholars of the real?

The book pokes, with an evidentiary prod, at our traditional self-notions. I aim to toss into the ring alternative viewpoints from which one might behold and experience the personal reality of self. Science—physical, biological, and psychological—has exposed tenuous impulses percolating in the subjective realm, and for me the weight of evidence has become overwhelming. I need to get over myself. The story of moi, like every story, begins with a Bang.

# THE COSMOS

*"Billions and Billions"*
*- Johnny Carson, then Saturday Night Live, impersonating astronomer Carl Sagan*

*"The size and age of the Cosmos are beyond ordinary human understanding."*
*- Carl Sagan*

*"In order to understand where we come from we have to understand events that happened in the first few seconds of the life of the universe."*
*- Brian Cox, particle physicist*

---

## *STARRY STARRY NIGHT*

Hop on the Harley. Let's scooter out of town on a moonless night, away from city lights, and look at the night sky. If the air is calm we'll see about two thousand stars, and we'll surely notice a pale white band spanning the sky, our galaxy, the Milky Way.

Earth is currently located near one of the five outward-spiraling arms that swirl away from the Milky Way's elongated disc-shaped core. As the Earth revolves around the Sun, so our entire solar system orbits the center of the Milky Way. Most scientists believe that what holds our pinwheeling galaxy together is the gravitational force of a supermassive black hole at its heart. Looking up at the night sky, our view is through and across the midsection of the galaxy, its discus shape displayed from Earth's perspective as an arching silvery ribbon. Every discrete star we can see is in the Milky Way galaxy, though most of the stars within our galaxy aren't visible to the naked eye. We can't, unaided, see individual stars within any other galaxies, but of course they're out there—stars, galaxies, and other celestial denizens—billions and billions of them.

Carl Sagan's final book, *Billions & Billions: Thoughts on Life and Death at the Brink of the Millennium*, starts out, "I never said it. Honest." From there, Sagan rattles off a bunch of monumental numbers that he admits he might really have used in describing the profound immensity of the universe. Among Sagan's wry justifications for disclaiming the phrase "billions and billions" is its imprecision. In fact, astronomers now believe there are at least 100 billion stars in our galaxy, and that there are, at minimum, 100 billion galaxies in the universe.

As we gaze into the starry night sky, it's natural to contemplate, as best we can, our place within the incomprehensible vastness of space and time. It's a humbling experience, though not for long. Could there be a discrepancy greater than the one between the microscopic speck in spacetime that, objectively, we know ourselves to be, and our subjective personal universe, where we find ourselves the centerpiece of existence?

## SIZE MATTERS

Size of moi: To the nearest meter, I'm two meters (six foot six inches) tall.

Size of Earth: Earth's diameter is 13,000,000 meters (8,000 miles), which means that if I were laid out end-to-end it would take 6,500,000 of me (which is at least 6,499,999 too many for comfort) to span Earth's diameter.

Size of the Solar System: The diameter of the solar system, which now extends only to Neptune instead of Pluto, is close to 10 trillion meters. (If you, like me, felt a bit betrayed when Pluto was somewhat arbitrarily demoted, you might enjoy a bumper sticker that appeared at the time: "They got Pluto. Uranus is next.") Our solar system is almost a million times wider than Earth, which is why building a scale model of the solar system requires plenty of room. If we were to place a one-inch marble (the Sun) at the goal line of a 100-yard football field, then Neptune would be a tiny grain of sand, tiny even by grain of sand standards, at the other goal line. Now, if the number 10 trillion—the diameter of our solar system in meters—seems astronomic, try not to think about the fact that the U.S. national debt is almost twice that figure, at $17 trillion and counting.

Size of the Milky Way: Our galaxy is about 250,000 light years in diameter, and one light year is approximately ten quadrillion meters, so the diameter of the Milky Way works out to two and a half sextillion meters, which is 250 million times wider than the solar system. A scale model of our galaxy, squashed down into a flat two dimensions, would occupy 500 quintillion football fields, laid out like huge grassy tiles, end-to-end and side-by-side. No doubt these numbers seem meaninglessly large, but that's the point.

Size of the Universe: We can only talk scientifically about the observable universe, which includes absolutely everything for which we have evidence of any kind. We can only "see" so far, and nobody knows what lies beyond. It seems likely that a complete universe, of whatever conformation, would be bigger than an observable one, but these are odd matters, and indeed one can find widely varying estimates even for the diameter of the observable universe. For our purposes, which is to put ourselves in our proper place, we'll use a conservative estimate of 27 billion light years. Once more converting light years to meters we reach the incontrovertible conclusion: universe big, me small.

Within our kaleidoscopic cosmos, distance isn't the sole elicitor of mind-jarring statistics. What follows are a few uncanny numerical facts, sans calculations or zeroes abreast, regarding neutron stars, one of the Earth's most attention-grabbing celestial cohorts.

Neutron stars are the densest objects known (because politicians aren't technically "objects"). Every star's lifespan is mediated by an astral tug-of-war between its outwardly propulsive fusion furnace and the compressive force of gravity, and how each star dies depends on its size. If a star is massive enough, when it runs out of nuclear fuel gravity does its victory dance. The star collapses on itself at up to one-fourth the speed of light, to a size as small as 30 kilometers (19 miles). All that's left is an ultra-condensed core, 100 million million million times as hard as a diamond.[6] At that point, the star's atomic nuclei are so crushed together that quantum mechanics shouts "no mas," and like a cosmic Superball the star's stellar matter rebounds, producing the most violent explosion we know of, a supernova. The resulting shock wave produces the highest temperatures in the universe, at over 100 billion degrees Kelvin. Even though neutron stars can be as small as 30 kilometers in diameter, they're more massive than the Sun. According to NASA's Goddard Space Flight Center, one teaspoonful of neutron star, on Earth, would weigh a billion tons.[7]

## BIG BANG

All cultures brandish creation myths. We dig ourselves, and we want to know from what great or puny fount we ascended. Legends get handed down from generation to generation, often with embellishment and frequently featuring political spin, but in truth nobody knows how the universe began, or if it necessarily had a beginning. There weren't any eyewitnesses, so scientific evidence is all we've got, circumstantial though it may be.

In 1980, when Carl Sagan wrote *The Cosmos*, controversy lingered regarding the origin of the universe. Some support continued for the Steady State model, which held that the universe had always existed more or less as it is now, but since then data have accumulated showing that the Steady State model is wrong, which is to say that calculations based on it were shown to be incorrect. On the other hand, predictions based on the Big Bang model keep turning out to be right, so almost all scientists now accept the Big Bang.

The Big Bang was inferred, in large part, by noting the expansion of the universe then extrapolating backwards to the point at which it contracted into its original state. The particulars of the Big Bang, as now set forth by science, are mind-boggling. The following account of the first moments of the Big Bang is consensus-based theory as transcribed from an account by au courant particle physicist Brian Cox, described by the *Los Angeles Times* as "Carl Sagan with a Britpop haircut."[8] As far as we know, all that exists, including you and me, commences right about now.

The first scientifically plausible timeframe, known as the Planck Era, began $10^{-43}$ seconds (technically known as "one New York second") after the Big Bang. At this point there was no matter, just energy and a "superforce," the undifferentiated precursor to the four forces of nature we know today—gravity and electromagnetism plus the strong and weak nuclear forces. At the end of the Planck Era, gravity was the first to separate from the superforce. The Grand Unification Era ended $10^{-36}$ seconds after the Big Bang, with the strong nuclear force splitting from the superforce, after which the universe underwent a violent inflation wherein it swelled to $10^{26}$ times its previous size in a matter of $10^{-32}$ seconds. Subatomic particles then came into existence, though none of them yet had any mass. (I'm not making this up.)

The next major event occurred $10^{-11}$ seconds after the Big Bang, and (according to Cox) this is where actual scientific evidence can be brought to bear, thanks to re-creations and observations at CERN's Large Hadron Collider. The final two forces, electromagnetism and the weak nuclear force, separated from the superforce, while quarks and electrons, the building blocks of all matter, acquired mass. This process is called the Higgs mechanism, and the search for the Higgs particle (sometimes called the Higgs boson, and in the media, the "God particle") has been the holy grail of particle physics.

---

### The Holy Grail of Particle Physics

I should point out that "the holy grail of particle physics" is not a microscopic chalice, nor would it confer eternal life if you were to drink from it utilizing nano-tweezers. The holy grail of particle physics is a metaphor. The phrase is meant to connote ultimate worth, in this case the specific importance of the Higgs boson to scientific research funding. The "God particle" has attracted money like a stripper's G-string, which is not merely a tasteless analogy, it's a simile. As for the real Holy Grail, most archaeologists believe that it was discovered by Harrison Ford in *Raiders of the Lost Ark*, but in truth it was tracked down six years earlier in England by Monty Python.

(*Actual Dialogue*)

| | |
|---|---|
| Galahad: | What language is this? |
| Bedevere: | Brother Maynard, you are a scholar. |
| Maynard: | It is Aramaic! |
| Galahad: | Of course. Joseph of Aramathea! |
| Arthur: | What does it say? |
| Maynard: | It reads, "Here may be found the last words of Joseph of Aramathea. He who is valorous and pure of heart may find the Holy Grail in the Aaarrrgghh ..." |
| Arthur: | What? |
| Maynard: | "The Aaarrrgghh ..." |
| Bedevere: | What's that? |
| Maynard: | He must have died while carving it. |
| Bedevere: | Oh, come on. |
| Maynard: | That's what it says. |
| Arthur: | But if he was dying, he wouldn't bother to carve "Aaarrrgghh." He'd just say it. |
| Maynard: | But it's engraved right there. |
| Galahad: | Perhaps he was dictating. |

At long last, about a millionth of a second after the Big Bang, the quarks had cooled off enough for the strong nuclear force to pull them together into protons and neutrons, and the simplest element, hydrogen, was formed, its nucleus consisting of a solitary proton. After three minutes the universe had cooled sufficiently for the second simplest element, helium, to form, along with traces of lithium and beryllium, and possibly boron, the next lightest elements. As spacetime continued its now-slower expansion, there existed the four forces of nature and gas clouds composed largely of hydrogen. It was from these few primordial raw materials that the universe mindlessly organized itself.

## SPACETIME

To support the kind of life that's founded on our earthly brand of chemistry a planet can't be too hot or too cold (we need liquid water), which is why scientists say that Earth is in a "Goldilocks Zone" relative to the Sun. To me, the Earth is also in a Goldilocks Zone with respect to size and speed.

---

### Goldilocks and the Three Shrinks

You remember Goldilocks. Not the pubescent blonde who hopped from bed to bed because she wanted it "just right." That would be too hot. And not the tiresome towhead who tediously touted tepidness. Sorry, that was cold. I'm talking about a small-town girl whose story psychologist Bruno Bettelheim described as, "a struggle to move past Oedipal issues and confront adolescent identity problems,"[9] whereupon fellow psychologist Alan Elms chimed in that Bettelheim, "may have missed the anal aspect of the tale that would make it helpful to the child's personality development."[10] Harvard professor Maria Tatar chided Bettelheim as well: "While the story may not solve Oedipal issues or sibling rivalry as Bettelheim believes 'Cinderella' does, it suggests the importance of respecting property and the consequences of just 'trying out' things that do not belong to you."[11] Are these people kidding? No wonder Goldilocks fled into the woods. (I'm surprised Cinderella didn't join her.)

---

My point is that on a celestial scale, the Earth and everything on it is of modest size—not too massive, not too minute—and that velocity-wise, we operate at a pace nowhere near the speed of light. Where mass and velocity are not extreme, the "normal" laws of physics, Newtonian mechanics, apply. Space, time, and motion adhere to relatively simple formulas, such as the first great law of physics, Newton's Law of Universal Gravitation.

$$F = G \frac{m_1 m_2}{r^2}$$   $m$ is mass, $r$ is distance, $G$ is a constant $(6.67 \times 10^{-11})$

Every object attracts every other, and their mutual gravitational attraction increases with the product of the two masses and decreases with the square of the distance between them. Newton used his formula to predict with virtually perfect accuracy the movement of the heavens, and the law is still used to guide space vehicles.[12] For example, it helped

Tom Hanks and the other Apollo astronauts rescue themselves by sling-shotting around the moon.

| *The Real Problem on Apollo 13* | |
|---|---|
| *(Made-up Dialogue)* | |
| Hanks: | Houston, we have a problem. |
| Houston: | What is it? |
| Hanks: | There are no women on this mission. |
| Houston: | Well solve Newton's equation and get home. You'll be rich and famous, and that should fix the problem. |
| Hanks: | Roger that! |

Newton's law isn't genuinely universal, though, because it fails at subatomic dimensions, where quantum mechanics is the most accurate model so far. Quantum mechanics is a cirque bizarre, but it isn't directly relevant to the story of moi, so there's not much about it here. What's more germane to the emergence of you and me is that Newton also fails writ large, in the presence of supermassive objects like black holes, and at speeds approaching light. Under those circumstances (and under Newtonian conditions, but not within the subatomic realm), Einstein's general theory of relativity has proven unerringly accurate.

For our story, the importance of relativity is that it demonstrates beyond doubt that time is not syncopated; time is not a metronomic pendulum of experience. If we put a high precision clock in a jet, the clock runs measurably faster than its duplicate model on Earth. The phenomenon is called time dilation, and if GPS systems didn't account for it they'd be inaccurate by as much as ten kilometers a day.[13]

## THE PARADOX OF TIME

Personal experience is sufficient to reveal the paradoxical nature of time. On the one hand our mirror tells a tale of (to be generous) time's passage, while on the other we never experience a single instant that isn't now.

Also, because of light's finite speed we never see anything in real time, only as it used to be. When we look in the mirror we see ourselves as we were nanoseconds ago, and on the cosmic scale we're typically viewing ancient history. What appears to us a star could in fact be long gone. If our sun were to instantly and completely vanish from space, it wouldn't be until eight minutes later that we'd realize something wasn't quite right (is that a new shirt?).

Another of time's quirky features is that the years keep getting shorter. Like most people, I remember my youthful summers as endless, but my only memory of last summer is that it must have been canceled. I was ready to blame Einstein, or a government conspiracy to accelerate tax dates, until computational scientist T.L. Freeman figured out what's going on (and the news isn't good). Simply put, each successive year comprises less of our existence. At age 10, for example, a year represents 10% of our lives, but by age 50 a year constitutes only 2%, which, not surprisingly, feels a fifth as long. According to Freeman's adroit albeit unsettling calculations, three-quarters of an individual's lifetime (from an internal perspective) has passed by age 30, and at 50 we've lived 87% of our perceived lives. It's later than we think during every year of our lives except one, the last, which is the point at which actual years catch up with felt years. Finally (in the literal sense), we're as old as we feel. Freeman's theory seemed clever until I thought about it.

Einstein too theorized about time. As he wrote to a colleague, "The past, present and future are only illusions, even if stubborn ones." Einstein's special relativity stipulates that two events that occur at the same moment when observed from one reference frame could occur at different moments if viewed from another. More generally, there's just nothing within known physics that corresponds to the passage of time.[14]

That said, time gets a bad rap anyway thanks to its asymmetry, its refusal to back up and allow do-overs. In 1927, British astrophysicist Sir Arthur Eddington coined the term "arrow of time" to indicate time's relentless advance. The arrow of time impacts a variety of disciplines, including thermodynamics, cosmology, biology, and psychology. In thermodynamics the arrow points to entropy and chaos, in cosmology to an expanding universe, in biology to death and decay, and in psychology to past and future, memory and anticipation. (Time, as we'll see, is a critical element in the development and maintenance of self, because what would "identity" mean in the absence of time?)

Eddington put it this way: "Let us draw an arrow arbitrarily. If as we follow the arrow we find more and more of the random element in the state of the world, then the arrow is pointing towards the future; if the random element decreases the arrow points towards the past."[15] If entropy is true, a scientific given, and the future is always more random and disorderly than the past, then what accounts for the amalgamation over time of complex and highly organized entities such as planets, living beings, and technology? That's the subject of the next chapter. The point for now is that time, whether viewed personally or scientifically, is more enigmatic than convention lets on.

## STARDUST MEMORIES

The Big Bang resulted in cosmic clouds of hydrogen atoms, along with some helium and traces of the next two or three lightest elements. That was the extent of the original Periodic Table.

An element is composed of one type of atom, and its number in the Periodic Table is determined by the number of protons in the atom's nucleus. If variations are found in the number of neutrons within the nucleus, the variants are referred to as isotopes of the element; if the number of protons changes it becomes a different element. Ninety-two naturally occurring elements, up to #92 uranium, have been found on Earth, but if only the four or five lightest elements emerged from the Big Bang, where did all the heavier elements come from? The short but showy answer is stellar nucleosynthesis, nuclear fusion reactions within stars that compact heavier elements from lighter ones.

A hydrogen atom has a single proton, helium two. The most common nuclear fusion reaction assembles helium from hydrogen by crushing two hydrogen nuclei together, doubling the proton count. In stars, the hugely compressive force responsible is gravity. Nuclear fusion reactions release excruciating energy, and like every live star, the Sun is a fusion furnace. The resultant energy fosters life on Earth, and that's the good news.

It's fusion again that accounts for the horrific power of a hydrogen bomb. Inside an H-bomb there's a central hydrogen core that contains trillions of deuterium and tritium atoms, and while the garden variety hydrogen nucleus contains one proton and zero neutrons, its isotopes deuterium and tritium have one and two neutrons respectively—more stuff to blow up. Small atomic bombs, which release energy via fission (the splitting of nuclei), surround the hydrogen core, and when the atomic bombs are triggered by TNT their explosive force compresses the hydrogen core to the point of sustained eruptive fusion.

Nuclear fission was the agent of death for a quarter of a million Japanese, but it's also what powers electric plants. Regarding fusion, there are thousands of hydrogen bombs stockpiled in the United States, Russia, the UK, France, China, and India, but not one of those countries, nor any other, has discovered a peaceful use for nuclear fusion. And that's the bad news.

## How Well Do You Know Your Nukes?

See if you can figure out what real or fictional characters were involved in the following dialogue, which takes place in the midst of the Cold War between a General and his adjutant. The adjutant has just warned the General that the entire U.S. fleet of B-52 bombers is vulnerable to attack.

*(Actual Dialogue)*

General:   If I see that the Russians are amassing their planes for an attack, I'm going to knock the shit out of them before they take off the ground.

Adjutant:   But General, that's not national policy.

General:   I don't care, it's my policy. That's what I'm going to do.

The year was 1957. The exchange took place between General Curtis LeMay, head of the Strategic Air Command, and Robert Sprague, chairman of the Gaither Committee, a top-secret U.S. defense panel. To me it sounds like something out of *Dr. Strangelove* (1964), but according to Fred Kaplan of *The New York Times*, director Stanley Kubrick knew nothing about the conversation between LeMay and Sprague. Kubrick nevertheless scripted similar scenes, as when General Jack D. Ripper (Sterling Hayden), who has just ordered U.S. bombers to attack the Soviet Union, informs his British adjutant Lionel Mandrake (Peter Sellers) that, "I can no longer sit back and allow Communist infiltration, Communist indoctrination, Communist subversion, and the international Communist conspiracy to sap and impurify all of our precious bodily fluids."

Dr. Strangelove himself later explains the demands that will be placed on his personal entourage of select men, along with ten times as many women, in order to repopulate the planet after nuclear war: "But it is, you know, a sacrifice required for the future of the human race. I hasten to add that since each man will be required to do prodigious service along these lines, the women will have to be selected for their sexual characteristics, which will have to be of a highly stimulating nature."

Despite its farcical elements, *Strangelove* got so many details right that when defense analyst Daniel Ellsberg (who later leaked the Pentagon Papers) exited the movie theater, he turned to his companion and said, "That was a documentary!"[16]

Throughout every star's lifetime, the expulsive energy of fusion engages in a nonstop sumo bout with the compressive force of gravity. Eventually, though, hydrogen fuel runs out and gravity prevails. As the star collapses its temperature climbs, triggering a more profound fusion that refuels the star's nuclear furnace. Helium nuclei meld together, eventually forming carbon, and then, in a stellar blink of the eye, helium fuses with carbon to produce oxygen. Two of the cornerstones of earthly biological existence, carbon and oxygen, have come into existence.

For an average star like the Sun alchemy ends there, but for more massive stars the show goes on. As helium fusion ends, gravity again gets the upper hand and collapse resumes, reinitiating the fusing of atomic nuclei and creating ever-heavier elements up to #26, iron. The star's core has been fused into iron, but seconds after gravity has raised its trophy the star can burst into a planetary nebula (delightful to view through a telescope), but where are the 66 heaviest elements?

The fate of the largest stars is the most dramatic. As they die, gravity contracts their supermassive iron core at up to one-fourth the speed of light, producing the aforementioned neutron star, which can quickly explode in a supernova. Although details are not entirely clear, it's virtually certain that within those few convulsive seconds supernovae fuse and blast into space all of the heavy elements. The elemental clouds loiter around the universe until gravity re-congeals them, sometimes into planets, from which may arise inhabitants. We are indeed made of stardust.

---

### Attention Chocolate Lovers

Toast was popular in 18th century Scotland, tortillas not so much. One would expect this to halve the count of cases in which the supernatural visage of some legendary being—let's call him Elvis—materialized on a flour canvas. As one might suspect, miracles were more fashionable in 18th century Scotland than they are today (for instance, in our era a mere three-quarters of Americans believe in angels). Scottish philosopher David Hume, in 1748, defined a miracle as an apparent transgression of natural law, like a dead person coming back to life, and declared that he would accept a miracle on one condition only: All other possible explanations were more outlandish than the purported miracle.

In 1998, scientists noted inexplicable observations coming from the Hubble Space Telescope. Till then, it had been assumed that while the universe was surely expanding, the attraction between celestial objects, gravity, would gradually slow its expansion. The Hubble data revealed instead that expansion was accelerating. According to NASA, "No one expected this, no one knew how to explain it. But something was causing it."[17]

The mysterious new agent is called dark energy, a name that has nothing to do with the blackness of space or the dark side of The Force, Luke. It's called dark energy because we're completely in the dark; nobody knows what it is. And by the way, dark energy is now believed to comprise 70% of the universe, with dark matter making up another 25%, and conventional matter—everything we know of—comprising less than 5%. (I'm still not making this up.)

It's pretty fantastical stuff, and on October 4, 2011, The Royal Swedish Academy of Sciences announced that the Nobel Prize in Physics for 2011 would go to Saul Perlmutter, Brian P. Schmidt, and Adam G. Riess, "for the discovery of the accelerating expansion of the universe through observations of distant supernovae."[18]

I won't second-guess the Nobel committee and their peer review process, but Hume might have a quibble. When I was a graduate student we had an expression for any adjustment made to unanticipated data in order to make it fit a model. We called it a fudge factor. Again according to NASA, "We know how much dark energy there is because we know how it affects the universe's expansion. Other than that, it is a complete mystery."[19] So dark energy must exist, whatever it is, because it reconciles what would otherwise be anomalous observations. Hume might ask, as have some scientists, whether there are other possible explanations that are less bizarre than a mystifying new force of nature.

Personally, I accept Hubble's novel data, and I see it as a stroke of cosmic good luck that three-fourths of the universe is composed of award-winning dark fudge. Recent brain imaging studies on the neural effects of chocolate suggest that that should help motivate women to take up astrophysics. [Fill in your own retort here.]

# PROBABILITY AND RANDOMNESS

*"Misunderstanding of probability may be the greatest of all impediments to scientific literacy."*
*- Stephen Jay Gould, evolutionary biologist*

*"Our brains are just not wired to do probability problems very well."*
*- Persi Diaconis, professor of statistics and mathematics*

*"You've got to ask yourself one question, 'Do I feel lucky?' Well, do ya punk?"*
*- Dirty Harry Callahan, fictional cop*

---

## HEADS I WIN, TAILS YOU LOSE

On Sunday, November 13, 2011, as the New Orleans Saints lined up to kick off to the Atlantic Falcons, NFL play-by-play man Kenny Albert informed viewers that, "The Saints are now zero for ten on opening coin tosses this season." You might be wondering about two things, so yes, Kenny Albert is the son of Marv Albert, and no, it wasn't the infamous replacement refs bungling the flip. But could unscrupulous referees rig the toss, or were the Saints just unlucky?

I refer to Persi Diaconis as a "mathemagician," a mathematician who also performs a magic act. He taught himself to flip a silver dollar to land heads or tails, then promptly incorporated this "psychic ability" into his show. If you think a talent like that might come in handy for a bar bet, Penn Jillette, of the magic team Penn and Teller, can explain what it takes. "The big secret of magic is we are willing to work harder to accomplish something stupid than you can imagine. We'll practice things for years that you wouldn't consider investing an hour in."[20]

It seems unlikely that referees spend years practicing the coin toss, so assuming the NFL uses fair coins the Saints were just unlucky. How unlucky depends on the question we ask.

Put broadly, "How often should we expect to see a team lose 10 consecutive coin flips?" The probability of losing the first toss is ½, then each successive toss halves the odds again, and we discover that the probability of losing 10 consecutive tosses ($0.5^{10}$) is close to 0.001, or one in a thousand. A 16-game season presents 7 possible streak durations, starting with games 1-10 and ending with games 7-16, which ups the odds by a factor of 7, and there are 32 teams, which increases the odds again by a factor of 32. The chance that some team would lose ten straight coin tosses turns out to be .224, nearly 25%, which means that one of the NFL teams should suffer a 10-game flip-losing fiasco every 4 years or so.

If we ask about the first 10 games of the season that cuts the possible intervals from 7 to 1, and we'd expect some team to lose the first 10 coin tosses about every 32 years. If we ask specifically about the New Orleans Saints that's 1/32 as likely, which works out to about once every thousand seasons. I'm planning to join Saints fans on Bourbon Street to celebrate the news.

Probability theory arose as a branch of mathematics in the mid-seventeenth century, before which, in order to make sense of a spate of misfortune, many turned (as they still do) to the grand old chestnut of agency, divine intervention. By that account the Saints' losing streak would be an omen, most likely a punishment for their blasphemous name.

---

### When Bad Things Happen to Good People

Bad things can happen to good people, like when a drunk woman kept wrecking Lindsay Lohan's car, but the title of "Divinity's Most Abused Mortal" has to go to Job, whose ill fortune has been chronicled in the Old Testament, the New Testament, the Quran, the Doctrine and Covenants of the Mormon Church, and the "Cartmanland" episode of *South Park*. ("All of Job's children are killed, and Michael Bay gets to keep making movies. There isn't a God.") The scriptures agree that Job was a good and faithful man, tested by God, yet when Job demanded that He explain Himself, the Almighty beat around the proverbial (but in this case not the burning) bush. God might as well have told Job exactly what he told Larry, not in the Old Testament but in the Old Joke. After a run of appalling luck, Larry humbly petitioned the Lord for an explanation, to which God replied: "I don't know Larry, there's just something about you that ticks me off."

---

We need to ask a final question then bid adieu to the Big Easy. Given their 10 previous losses, what are the odds that the Saints would lose the coin toss at the start of their 11$^{th}$ game? If you gave an answer other than 50-50, 50%, 0.5, half, or even, you just fell into the Gambler's Fallacy, an erroneous notion that luck, after deviating in one direction, will then deviate in the other. As a matter of fact, in their next game on Monday night against the Giants, the Saints won the opening coin toss. They were due! (I assume you didn't fall for that one.)

## GALTON'S PARADOX

Sir Francis Galton, knighted in 1909, was an inventor, statistician, meteorologist, tropical explorer, and anthropologist (he also happened to be a half cousin of Charles Darwin). Among other accomplishments, he formulated the statistical concept of correlation, devised the first weather map, pioneered the use of surveys and questionnaires for data collection, and coined the term "nature versus nurture." Galton also led research related to the power of prayer, concluding that it had none since he could find no effect on the longevity of those prayed for.

He posed "Galton's Paradox" in the February 15, 1894 edition of the scientific journal *Nature*. "If you throw three pennies on the ground, what is the likelihood that all three will turn up the same?" (Three heads or three tails.) Galton's solution: Since it's 100% certain that two coins must turn up the same, the result will be determined by a third coin, which has a 50-50 chance of showing the same side as the other two. The odds, therefore, of three tossed pennies showing an identical surface are one in two, which means that three of a kind should ensue on half of all throws.

Really, Sir Francis? An experimentalist might test the probability empirically, throwing three coins a number of times, noting the results, then calculating the proportion of threesomes. Galton wrote that he did this, but conveniently failed to specify the results of his experimentation. Simpler and more accurate for problems that have a limited number of potential outcomes is to prepare a list that includes the possibilities, which in this case are HHH, TTT, THH, TTH, HTT, HHT, HTH, and THT. By inspection, the odds of throwing three of a kind are 2 out of 8, or once every four tosses.

Which solution is correct? Galton's logic appears faultless, but the enumerated outcomes say otherwise. That's why it's a paradox, the resolution of which could surely be ascertained by throwing some coins.

## PHYSICIAN, HEAL THYSELF

Galton posed his three-coin "paradox" because he thought it an entertaining reflection on the deceptiveness of probabilistic reasoning. The correct solution, as we might imagine, is one in four throws. As clarified by Galton himself, the question isn't about the probability of how *a* third coin (an independent third coin) might land, which would be 50-50, it's about how *the* third coin could land in conjunction with two others, and that can be examined most concisely via the roster of potential outcomes.

| OUTCOME | FIRST 2 COINS | "THE" 3rd COIN |
|---------|---------------|----------------|
| HHH | HH | H |
| TTT | TT | T |
| THH | TH | H |
| TTH | TT | H |
| HTT | HT | T |
| HHT | HH | T |
| HTH | HT | H |
| THT | TH | T |

If Galton's brainteaser leaves you a bit doubtful, you can begin to appreciate the innate trickiness of probabilistic thinking. Galton, of course, was showcasing a diversion for his science-geek contemporaries; not as trifling are results from a study at a prominent medical school where physicians, residents, and fourth-year students were asked:

> If a screening test to detect a disease, whose prevalence is known to be one in a thousand people, has a false positive rate of 5%, what is the chance that a person found to have a positive result actually has the disease (assuming you know nothing about the person's symptoms or signs)?

Take a minute to review the question, since most of us eventually face such a health assessment. And note that medical tests are typically designed to be overly sensitive, producing false positives, because a less reactive test would yield more false negatives, where a patient who has the disease would go undetected.

If your arithmetic suggests that given a positive test result there's a 95% likelihood that the patient has the disease, you're in agreement with roughly half of the physicians, residents, and fourth-year medical students. Only eighteen percent of them arrived at the correct solution, 2%.

Suppose that we test 1000 people. The prevalence of the disease is known to be 1 in 1000, hence we'd expect 1 true positive result. The test will produce a false positive 5% of the time, so approximately 50 of the remaining 999 people will turn up positive. Of the 51 expected positive results only one person will actually have the illness, which means that the odds of having the disease, given a positive test result, would be 1 in 51, or 2%.[21]

The bad news is that the intrinsic slipperiness of probabilistic reasoning foiled half the medical professionals, within a scenario they regularly encounter. The good news is that a positive screening might not be as dire as it feels when you first get the results. And here I speak from personal experience. While consulting for the Infectious Diseases Division of the New Mexico State Health Department, I tested positive for tuberculosis during a routine screening. TB skin tests have a false positive rate of around 8%[22], and by luck my lot ultimately fell within that fortuitous sliver.

## LET'S MAKE A DEAL

Do you remember Monty Hall? If you said "I do," then by the psychic power vested in me by the state of self-delusion I now pronounce you old. Even if you don't remember Monty, full or in part, you've no doubt heard of his TV show *Let's Make a Deal*. With whatever's left of our probability-boggled minds, it's time to face the Monty Hall Problem.

Setup: You're the contestant, and Monty shows you three large doors. Behind one of the doors is a Lamborghini and behind each of the other two doors is a goat; you'll win whichever prize is behind the door you select. Monty asks you to pick one of the three doors and you do so.

Before opening your door, and with models pirouetting around like big-hair ballerinas, Monty opens one of the other two doors, one that he knows conceals a goat. (You can see it's a good-looking goat, but you'd still prefer the car.) Before opening your original choice Monty asks you, "Would you like to switch to the other door?" Should you?

Seems straightforward. The car is behind one of the two closed doors, your original door and the other unopened door, so it's a 50-50 proposition and there's no point in swapping. But, need I tell you, not even close! If you switch you double your chances of winning the car. If you're after a goat then don't switch.

The problem was originally posed by statistician Steve Selvin in 1975, then published fifteen years later in *Parade* magazine. Over 10,000 *Parade* readers, including nearly 1,000 with Ph.D.s, wrote in disputing the solution. You can find dozens of explanations on the Internet and elsewhere, each trying to make the unintuitive comprehensible. My own account will be brief, but I'll also tell you how I proved to my cynical self that it's absolutely correct.

Whatever door you initially pick, there's a 1/3 chance the car is behind it. There's a 2/3 chance the car is behind one of the other two doors. Monty doesn't open your original choice, but he opens one of the remaining two doors, one that hides a goat. If the car happened to be behind one of the two doors you didn't initially pick (a 2/3 likelihood),

Monty has shown you which of the two is hiding the car by eliminating the goat door. If you switch you'll win the car two out of three times, and if you don't you'll have your original 1/3 odds.

I found the solution difficult to accept so I decided to perform an experiment. I took three squatty candles that were identical except that two were tan (the goats) and one was red (the car). I put the candles on a table and with eyes closed scuffled them around until I lost track of where the "car" was, then from the resulting lineup chose the rightmost candle each time (it makes no difference which one you pick). If I stuck with that choice I'd be correct one-third of the time, and two-thirds of the time the car would be in one of the other two positions. Since Monty invariably eliminates the wrong choice of those remaining two, the last candle, the "other door," would win the car two out of three times. In essence, I would be doubling my odds by selecting two doors and letting Monty rule out the wrong one.

If it still seems odd try the experiment. You can perform it with any three identical objects, one of which represents the car. Pick the same door/object each time (it has no effect on the outcome), tally the results of sticking or switching, and you'll quickly see the 2 to 1 pattern emerge. Plus, it will be obvious why it must be so.

Try it for yourself—science works. Too lazy? Then ship me a Lamborghini and two goats. I'll perform the experiment myself and email you the results, along with a photo of a glass of fresh goat's milk in the cup holder of my new ride.

### THAT'S SO RANDOM

A scientist would tell us that the random and the improbable are fundamental attributes of many (some would say all) of the milestone events that culminated in you and me. The mindless sculptor we call evolution has fashioned our brains to identify patterns rather than the unordered, which is why human beings detect designs in the night sky, within clouds, and on a humble tortilla. Research shows we're poor at probabilistic reasoning and poor at recognizing the random. The truth, though, is that to appreciate how each of us came to be, we need a basic grasp of probability and randomness.

The table below displays a random sampling of 300 random digits from the 1955 page-turner *A Million Random Digits with 100,000 Normal Deviates*, a book you could buy today on Amazon for $65. The Rand Corporation has hauled in a truckload of money over the years for an assortment of keystrokes that any one of my cats could type.

| | | | |
|---|---|---|---|
| 73735 | 45963 | 78134 | 63873 |
| 02965 | 58303 | 90708 | 20025 |
| 98859 | 23851 | 27965 | 62394 |
| 33666 | 62570 | 64775 | 78428 |
| 81666 | 26440 | 20422 | 05720 |
| | | | |
| 15838 | 47174 | 76866 | 14330 |
| 89793 | 34378 | 08730 | 56522 |
| 78155 | 22466 | 81978 | 57323 |
| 16381 | 66207 | 11698 | 99314 |
| 75002 | 80827 | 53867 | 37797 |
| | | | |
| 99982 | 27601 | 62686 | 44711 |
| 84543 | 87442 | 50033 | 14021 |
| 77757 | 54043 | 46176 | 42391 |
| 80871 | 32792 | 87989 | 72248 |
| 30500 | 28220 | 12444 | 71840 |

As a matter of fact, if you find yourself in need of some random numbers it would be better to set your cat loose on the numeric keypad than to concoct them yourself. Our human inclination is to stagger the ten candidates evenly, varying the order and rarely including a repeater. That isn't true in the random number table above, where I've highlighted the five-digit sequences within which a single numeral appears three times or in which there's a back-to-back succession of a two-digit string (there are numerous unmarked two-time repeaters as well).

When Apple first introduced the shuffle feature on its iPods the shuffle was truly random; each song was equally likely to play at any time. But as soon as people noticed a song repeat they imagined a secret pattern, so Apple revised the algorithm. "We made it less random to make it feel more random," said Apple CEO Steve Jobs.[23]

---

### Rock, Paper, and Scissors

Question: Is rock, paper, and scissors a game of chance or a game of skill? Answer: It's a game of fortune, by which I mean it's worth a mint. Competitions now offer large cash prizes, and the winners are consistently better, across tournaments, than other players. Somehow, top competitors can anticipate opponents' moves, perhaps thanks to "tells" (behavioral giveaways), and certainly because contestants exhibit patterns. If you want to disguise your next throw, the best and only bet is to base your throws on random numbers. So next time you play rock, paper, and scissors bring along a random number table, or a cat, or better yet bring along an iPhone and one of the several hundred iPhone random number apps. Tip: If you plan to throw "paper," don't hold the phone in your throwing hand.

As philosopher Daniel Dennett notes, randomization is particularly important if you intend to play rock, paper, and scissors with God for your salvation, because He knows your mind. In the end (literally), what you'd like to hear Him say is, "That was so random, but you win and you're in."

Rock, paper, and scissors represents a trivial domain in which one might gainfully apply my new invention, Randomly Generated Behavior (RGB™), because sometimes it pays to go random. Take relationships for example. According to no less an authority than *Urbanette* magazine, "The longer you remain in a relationship with a significant other, the more likely you are to hit a point where he or she will take you for granted."[24] Unfortunately, even when we recognize our own staleness, habitual behavior patterns are difficult to change, and that's exactly where RGB™ comes in. For instance, decide which one of the following three scenarios would most benefit your relationship.

Typical: You walk into a party with your significant other, scout out your buddy or BFF, and start shooting the bull. From across the room your significant other waves that you should, "mingle, mingle."

Nice Try: You walk into a party with your significant other, pull up a random number on your iPhone, then approach the person whose clockwise position in the room matches the number. By chance, that person turns out to be spectacular, so of course you start shooting the bull. Nice try, but from across the room your significant other wags a big thumbs down.

Bingo: You walk into a party with your significant other, pull out your iPhone and generate a series of random numbers, then follow the list to chat briefly and cordially with each partygoer. From across the room your significant other blows you a promising kiss. RGB™ does its job, your relationship prospers, and I get my picture on the cover of *Urbanette* magazine.

Random numbers are important, but why? And what does the term "random" actually mean? First, there's no simple definition of what constitutes randomness; it's a bone of scientific contention. Having said that, most scientists and entrepreneurs are content to accept a pragmatic view of the random—are the items jumbled enough for the purpose at hand? Second, random numbers are big business. For one thing, they're the core technology behind data encryption. Third, from a religious, philosophical, or spiritual perspective, questions of free will, causation, determinism, and personal responsibility hinge in one way or another on the notion of randomness. Finally, and as mentioned, an awareness of probability and the random is essential to the story of moi.

For our purposes, let's live with the idea that randomness involves disorder and unpredictability. There isn't a discernible pattern within existing items, and we can't predict the next item. When I'm asked to randomly select people for some task or privilege, I typically put their names in a hat, shake the hat, then draw names. Is it random? It might be but it's not guaranteed, and there's no way to prove it one way or the other.

Consider the initial Selective Service Draft Lottery in 1969. All American men born between 1944 and 1950 participated in the drawing, and the prize was likely service in Vietnam. The men would be sorted for service by birthday, those with the first-drawn birthdays going in first. The numbers 1 through 366, for the days of the year, were written on strips of paper and each slip was placed in a separate plastic capsule. The capsules were placed in a shoebox on a month-by-month basis, where they were mixed and then dumped into a deep glass jar. Capsules were drawn from the jar one at a time as young men across the country cringed in front of television sets.

It quickly became clear that birthdays occurring later in the year, particularly in November and December, were being first-drawn with unlikely frequency. Subsequent analysis concluded that the shoebox shuffling may have been insufficient to disorder the original month-by-month insertion pattern.[25] Despite general agreement that the drawing was at best nonuniform, the lottery was allowed to stand. It cost some their lives, but as James Shirley wrote, "there is no armor against fate."

## MONKEY SEE, MONKEY TYPE

Here's how Persi Diaconis (the psychic mathemagician) and his Harvard colleague Frederick Mosteller summed up the law of truly large numbers: "With a large enough sample, any outrageous thing is apt to happen." *The New York Times* titled the professors' story "1-in-a-Trillion Coincidence, You Say? Not Really, Experts Find." Drs. Diaconis and Mosteller analyzed ten years of data related to eerie coincidences, and showed that many events that looked extremely unlikely were almost to be expected. *The Times* cited an example of a woman who won the New Jersey Lottery twice in four months, which had been widely reported as beating seventeen trillion to one odds, but careful analysis showed that the chance of such an event happening *to someone somewhere in the United States* was more like one in thirty.

Impossible events never occur, but improbable events, no matter how unlikely, are bound to happen by chance given exactly what the universe dishes up, immeasurable opportunity. "Nearly impossible" is another way of saying "sure to take place sooner or later," and make no mistake—you and I are highly improbable happenings.

---

### You Are a Rare Bird Indeed

I mean that in the good way, and in fact scientists have calculated how rare a bird you are. Dr. Ali Binazir looked at the odds of your parents meeting, the chances of them talking, forming a long-term relationship, having kids together, and of the right egg and sperm combining to make you. He went further back and looked at the probability of all your ancestors successfully mating, of all the right sperm meeting all the right eggs in each of your ancestors. His final tally aligned with previous scientific estimates that set the odds of a singular "you" at one in 400 trillion.

In a Buddhist parable, a solitary life preserver is thrown into an ocean where a sole turtle lives. According to the analogy, the likelihood of your unique existence is the same as that of the turtle poking its head out of the water into the middle of the life preserver, by chance, on one try. When Binazir computed oceanic volume, turtle dimensions, and life preserver whereabouts, his arithmetic put the odds of a terrapin bull's-eye at one in 700 trillion. He wrote, "The two numbers are pretty darn close for such a far-fetched notion, from two completely different sources, old-time Buddhist scholars and present-day scientists."[26] In other words, if your sweetheart tells you that you're one in a million it's not really a compliment—it's faint praise by a factor of 500,000,000.

---

I was taught in middle school that given an infinite number of monkeys and an infinite number of typewriters, the monkeys would eventually produce all the great works of literature. I'm still not certain if that's true, but I now know that the Infinite Monkey Theorem has been around for a long time, even as a publisher's business model. In a

*Simpsons* episode, Mr. Burns chains a thousand monkeys to typewriters, tasked with writing a great novel. Burns doesn't take it well when one of the monkeys types, "It was the best of times. It was the blurst of times."

Aristotle was probably the first to contemplate the implications of infinity (minus typewriters or *The Simpsons*). He believed that we can only think of the "potential infinite," because the "actual infinite" is simply not humanly imaginable. Weighing in on the topic over time were Cicero, Pascal, Jonathan Swift, and Sir Arthur Eddington (of time's arrow), who prophesized that, "If an army of monkeys were strumming on typewriters they might write all the books in the British Museum."

Reality Check: In 2003, students from the University of Plymouth put the notion of key-flogging simians to an empirical test by placing six crested macaques and a computer in a cage for a month. At the end of the experiment the monkeys had produced about five pages of letters, mostly "S," but not a single word. According to Mike Phillips, one of the study's researchers, the lead male spent most of his time bashing the keyboard with a rock, while the others urinated and defecated on it.[27]

Clearly, animal husbandry issues perturb the Infinite Monkey Theorem, plus there are mathematical and philosophical concerns as well, but the main point for us, again, is that within the immensity of time and space, low probability events, up to and including you and me, will eventually come to pass. But in our case infinity is off the table— scientific consensus stipulates that the universe sprang forth a meager fourteen billion years ago. To engineer us, the Howlers of happenstance were allotted fewer typewriters, which is why our lives read more like pulp fiction than great works of literature.

## FINAL EXAM

Question: If you randomly draw five cards from a deck, which of the following two hands would you be more likely to draw?

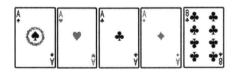 

Answer: Each is equally likely. There are 2,598,960 prospective poker hands, and the chance of randomly drawing any specific one of them is the same, 1 in 2,598,960. Poker-wise, the odds of drawing four aces is 1 in 54,145, whereas the odds of drawing "nothing" is 1 in 2, and while that's meaningful in a poker game, it wasn't the question. Every combination of five cards is equally likely; we just pay more attention to some patterns than others.

That's also why we fall for the "full moon fallacy," an urban myth that people are affected, psychologically and behaviorally, by phases of the moon. Dozens of studies have demonstrated there's no relationship between the moon and our mental states or behavior, including one large-scale study that examined three years of ER data and found that although 80% of nurses and 64% of doctors believed they had seen more psychogenic cases during the full moon, they hadn't. The researchers hypothesized that when patients show up on the day of a full moon, personnel think "ah, full moon," but on the other 28 days nobody thinks "ah, not full moon." The hospital staff's mental accounting fell prey to confirmation bias, the tendency to see what we expect, and to the base rate fallacy, where we pay excess attention to particular cases without regard to the baseline frequency of such cases overall.

Probability is tricky. If you flunked the final exam you're clearly an aficionado who sees playing cards in poker-vision, and if you passed, your middle name is clearly Poindexter (and congratulations).

## WHY THINGS CHANGE

The first law of thermodynamics is the conservation of energy, which states that energy can change from one form to another, but can't be created or destroyed. The second law of thermodynamics, regarded as one of the most profound laws of nature, describes the tendency of closed systems to dissipate into disorder; matter and energy are prone to disperse in disarray. Unless constrained, matter crumbles and energy spreads out.

It's entropy. It's the reason for time's arrow. It's why you, me, the Earth, the Sun, the Milky Way, and (most believe) the universe must die. It's a naive process, mindless, unmotivated, purposeless. It's the only thing guaranteed to make the sad happy and the happy sad, the wise man's one truth; this too shall pass.

In seeming contradiction, we see sublime organization everywhere we look, especially in the mirror. But let's not forget how digits repeat themselves in a table of random numbers, or Diaconis's law of truly large numbers, or Pulitzer Prize-winning simians. Sometimes, in the overriding process of dissipation, matter and energy find themselves slung together into a temporarily stable configuration.

Chemist Peter Atkins describes the exquisite forms into which the energy of random diffusion might organize itself. "If the spreading is captured in an engine, then bricks may be hoisted to build a cathedral. If the spreading occurs in a seed, then molecules may be hoisted to build an orchid. If the spreading occurs inside your body, then electrical and molecular currents in your brain may be organized into an opinion."[28]

For Atkins, all change, including human activity, is the simple unfolding of an aimless universal process. He acknowledges the sense of personal agency; inadvertent mechanisms have wired our brains to feel that. What Atkins sees as error, if not hubris, is the human inclination to project intention onto the universe at large.

The sole certainty is change. In a way, all things have a cause in that they result, more or less, from prior events, but being caused is not the same as having purpose or meaning. And though human beings delight in patterns, what seems a pattern can be random, and in any case a pattern is not a design. Wishful thinking, no matter how heartfelt, well-intentioned, or commonplace, does not constitute proof of that which is wished for. Each of us, like all that exists, surfs briefly the wave of unmotivated decay. We strive to hold ourselves together, which is why we must eat (and why we take yoga classes), but within a cosmic eyeblink we fall back to dust. There are two ways to take the news: That sucks, or I'm free.

# ORIGIN OF LIFE

*"Life results from the non-random survival of randomly varying replicators."*
*- Richard Dawkins, evolutionary biologist*

*"If God didn't do it this way, He missed a good bet."*
*- Harold Urey, Nobel chemist*

*"There's plenty of unearthly looking things moving around in my refrigerator, so there's always a chance of life springing up almost anywhere."*
*- Pete Conrad, astronaut*

---

### *SOMETHING FROM NOTHING*

Eddington put forth that as we trace the arrow of time rearward into the past we encounter less randomness, diminished entropy, more order—the second law of thermodynamics in reverse gear. As we wind our video of the expanding universe backwards to its first frame, we must arrive at a state of minimum entropy, maximum orderliness. That, fundamentally, is what scientists believe.

We've got tantalizing tidbits of evidence regarding events that followed within nanoseconds of the Big Bang, but nobody knows what happened at the pinpoint of inception. One of the open questions of cosmology remains: Given that the entropy of the universe has only increased, how did it get such a low entropy when it came into being?

Peering out at the universe now, we behold a troupe of glittery celestial dervishes whirling swiftly away from each other. So if existence happened to flick on as some thingamajig that was (compared to the current universe) smaller, more compact, or nonexistent, maybe the austere tidiness of that whatchamacallit reconciles with the second law. An alternate view is that natural laws simply didn't apply at the distinct moment of creation. There are those who say, and I'm ill-equipped to endorse or refute them, that both relativity and quantum mechanics break down in describing the Big Bang.

An even more provocative issue, perhaps, is whether something can spring from nothing. We don't know. We're also clueless as to *why* the universe began and as to what, if anything, might have come before it. Or if there even was a before. There are unanswered questions, but take note you golden dreamers, love-at-first-sighters, Scientologists, and other I-know-what-I-knowers: Science never stops edging cautiously forward, treading solely on a footing of evidence. So let's forget now about what we don't know and return to real life.

## LIFE AS WE KNOW IT

What is life? Based on what attributes do we decide if something is alive? Is there an unambiguous litmus test? How about an application process?

> ### The Facts of Life
>
> On an episode of *Sesame Street*, the late great Robin Williams disclosed to kids that to determine if something is alive you need to ask three questions. Does it eat? Does it breathe? Does it grow? He then removed his shoe and performed "experiments" to ascertain whether the shoe could indeed meet those criteria. It could not. Williams put the shoe back on and limped off stage right, muttering opaquely, "I'm tellin' ya, try a banana in your shoe and you'll know what a day is."[29]
>
> Well he might know what a day is, but Williams was impishly awry vis-a-vis the benchmarks of life. The notion of what constitutes life isn't simply a scientific concern—ramifications spill over into ethics, medicine, law, literature, philosophy, and film.
>
> You've probably watched at least one of the Terminator movies. Within the franchise storyline, the viral computer program named Skynet "begins to learn at a geometric rate. It becomes self-aware at 2:14 a.m. Eastern time, August 29th."[30] If something can replicate, can learn, and is self-aware, could it *not* be alive? And what about the female replicant, Rachael, in *Blade Runner*? Harrison Ford had *sex* with her! We can probably draw the line at "Embraceable Ewe," an inflatable plastic sheep that one of my buddies got at his bachelor party. A mule can learn, grudgingly one supposes, but it can't reproduce because mules are sterile. Is a mule self-aware? Beats me, but I'd bet it's alive.

Scientists, not surprisingly, have come up with a list of the seven characteristics of life. Living things: 1) are composed of cells, 2) require energy, 3) reproduce, 4) display heredity, 5) respond to the environment, 6) maintain homeostasis (a state of internal equilibrium), and 7) evolve and adapt. Does a mule make the cut now? True, it can't replicate, but it is itself the product of sexual reproduction, so yes, of course a mule is alive. Just FYI, a mule is a cross between a donkey stallion, called a jack, and a horse mare, while hinnies are the reverse, a stallion horse crossed with a donkey jennet.

Bacteria, like other single-celled organisms, meet every one of life's seven requisites, and that includes anaerobic bacteria, which, as I'm sure Robin Williams knew, don't breathe. Tinier still is the virus. In 1959, German scientist Wolfhard Weidel wrote a book called (get ready for an ingenious title) *Virus*, in which he avowed that a virus is, "midway between brute matter and living organism." The book's cover proclaims, "Nothing brings us so close to the riddle of life—and to its solution—as viruses."[31] Scientifically speaking that may be true, but I prefer to think of life not as a riddle to be solved, but as a mystery to be lived. On a less philosophical note, each of us has had to endure more than one up-close and personal confrontation with a virus.

For the last hundred years or so, science has flip-flopped back and forth as to whether a virus is alive. Although viruses exhibit heredity via DNA or RNA, and meet five of the other six criteria of life, they have no cells. Instead, viruses parasitize host cells, such as yours and mine, for raw materials and energy so that they can carry on with their miserable little lives, and reproduce. The fact that viruses are able to adapt to their environment makes them all the more dangerous and all the less lovable, plus viruses are quite amenable to genetic engineering, uncaring of its purpose, and they don't play political favorites.

How then, after all, should we distinguish life? Personally, I'm content to go with an analog of Supreme Court Justice Potter Stewart's well-known pronouncement, sometimes wrongly attributed to Oliver Wendell Holmes, during a 1964 obscenity case. Addressing hardcore pornography, Stewart famously agreed that though it was hard to define, "I know it when I see it."

### GET A LIFE

There's no doubt that at some point, on Earth, matter managed to stumble over two critical thresholds. The first was from inorganic gunk to organic goo, and the second was from organic goo to living glop. You and I are proof enough. But as at the cosmic inauguration, no paparazzi were on hand, so we can only reconstruct the origin of life from available evidence and via informed speculation.

The question some people ask is: What are the odds of something as subtle, complex, and downright extraordinary as life springing up, by sheer accident, on Earth? They answer: So infinitesimal that an agent— let's call him Elvis—must have had a hand in it. (Forgetting about how the agent himself originated, we might more importantly ask whether he's going to try to sell us primordial swampland.)

Professors Diaconis and Mosteller would say they're asking the wrong question. The real question is whether life could arise on any planet anywhere in the universe, and that ups the chances incalculably. Wherever that might be, that's where we would find ourselves. That's where we'd set up shop and start peddling our wares—rustic, exotic, artsy, pragmatic, puritanical, libertarian—each of us touting our own.

Whatever the odds life clearly arrived on the scene, but how, in Earth's 4.5 billion year history, did brute matter jump the fence? Though scientists quibble about the details, there's general agreement as to the overall progression of events, and a number of scientists have articulately tendered their renditions. For our narrative I'm going to rely primarily on Richard Dawkins's speculative account from *The Selfish Gene*, written in 1976 then upgraded in 2006 to the 30th Anniversary Edition.

---

### An Apology in Advance

I apologize for reducing Dawkins's eloquent prose to my less adroit "Cliffs Notes" version, but for our story compaction is the only option. According to Dawkins, "A neutron walks into a bar and asks, 'how much for a beer?' The bartender, a proton, replies, 'for you, *no charge*.' The neutron says, 'are you sure?' The proton replies, '*I'm positive*.'" But of course that's not Dawkins, and it's a bad joke to boot, because all the good chemistry jokes Argon. (Sorry, but I did apologize in advance.)

---

We know that Earth and its 92 natural elements coalesced from cosmic dust, along with the rest of our solar system, about 4.5 billion years ago. We have fossil evidence of life reaching back 3.5 billion years, but we don't know what chemical raw materials were abundant in the billion-year interlude between Earth's inception and the origin of life. Reasonable possibilities include water, carbon dioxide, ammonia, and methane, each of which has been identified on at least one of Earth's neighboring planets.

The genesis of life began when inorganic molecules within our atmosphere, sparked on by lightning and sunlight, vaulted threshold number one to become organic. The newly minted organic molecules, along with the usual lineup of inorganic suspects, floated about Earth's seas in a primordial soup.

Again energized by sunlight and lightning, the organic molecules combined themselves into larger ones. Today, large organic molecules would be quickly broken down by bacteria and other biotic late-comers, but back then, as Dawkins puts it, "large organic molecules were free to drift unmolested in the thickening broth."

It's worth interjecting that researchers have performed laboratory experiments attempting to refabricate the recipe for Earth's primeval soup, and although no menial life form has ever materialized, not even [fill in the name of one of your exes here], life's precursor molecules always do. In 1952, Miller and Urey performed an experiment in which a mixture of water, hydrogen, methane, and ammonia was cycled through an apparatus that delivered electrical sparks to the mixture. After a week they found that almost 15% of the carbon in the system was now in the form of organic compounds, including 23 amino acids, the building blocks of proteins.[32] Some scientists have proposed alternatives to the primordial soup theory, but they all lead to our second threshold: How did organic molecules come to life?

By happenstance, a remarkable molecule formed that was able to make copies of itself, which is exceedingly unlikely, but as decreed by the law of truly large numbers, outrageous events must transpire. Dawkins refers to this molecule as the "replicator," a term that has been broadly adopted, and he goes on to say that perhaps such a development is not as improbable as it might seem, and that it only had to happen once.

Let's imagine a large molecule as a chain of smaller building block molecules, and that building blocks were available in the soup, then let's further suppose that each building block had some type of affinity for its own kind, could nestle comfortably, atomically speaking, alongside a partner. As building block molecules drifted by they might sidle up to their matches within the larger molecule, eventually assembling a side-by-side facsimile of the original. This could have occurred many times before an accurate doppelganger happened to break off intact, now free and chemically apt to piece together additional clones.

But no copy process is perfect. As the initial replicator cast off twins some would have minor variances, and the soup would become populated by several varieties of replicators, all "descended" from the original. Some would, by chance, find more building blocks available, and some might be more proficient at assimilating neighbor molecules. Others might be more effective at sloughing copies, and some, who were less disposed to break apart, might have had more time to, shall we say, play the field.

Eventually, the soup would have become populated by the most stable of the replicators, stable in that individual molecules lasted a long time, or replicated rapidly, or replicated accurately. Those would be the molecules that predominated following competition for the increasingly in-demand building blocks.

For Dawkins, it's moot whether we label the replicators "alive" or not (at that moment in time, Robin Williams was unavailable to perform an experiment). The point is that as soon as replication depended on successful competition for resources, evolution had begun.

## SURVIVAL MACHINES

The first replicator strains competed for existence without hard feelings, but if a copying error produced a higher level of stability or a new way of impeding rivals, the change was preserved and it multiplied. Dawkins tells us:

> The process of improvement was cumulative. Ways of increasing stability and of decreasing rivals' stability became more elaborate and more efficient. Some of them may even have 'discovered' how to break up molecules of rival varieties chemically, and to use the building blocks so released for making their own copies. These proto-carnivores simultaneously obtained food and removed competing rivals. Other replicators perhaps discovered how to protect themselves, either chemically, or by building a physical wall of protein around themselves. This may have been how the first living cells appeared. Replicators began not merely to exist, but to construct for themselves containers, vehicles for their continued existence. The replicators that survived were the ones that built *survival machines* for themselves to live in. The first survival machines probably consisted of nothing more than a protective coat. But making a living got steadily harder as new rivals arose with better and more effective survival machines. Survival machines got bigger and more elaborate, and the process was cumulative and progressive.[33]

That was four billion years ago. The replicators didn't die out, they evolved. The process that we now call natural selection spurred an acceleration in the complexity and sophistication of survival machines. Today, the replicators swarm in huge colonies within every living thing. "They are in you and in me; they created us, body and mind; and their preservation is the ultimate rationale for our existence. They have come a long way, those replicators. Now they go by the name of genes, and we are their survival machines."[34]

As science writer (and member of the UK's House of Lords) Matt Ridley sees it, bodies don't replicate themselves, they're grown under genetic direction, whereas genes manufacture authentic copies, hence it follows that the body is an evolutionary vehicle for the gene rather than vice versa. If genes induce their bodies to eat, have sex, rear children, and fend off rivals, then the genes are perpetuated.[35]

### THE TRUE IMMORTALS

Ponce de Leon crisscrossed Florida in search of the fountain of youth, but data from the 2010 U.S. Census suggest he might have been looking in the wrong place—Florida now boasts the country's highest percentage of residents over 65. Or perhaps Florida's is a metaphoric fount, a wellspring that bestows not chronological youth, but youngness of heart. (We'd have to ask their cardiologists.) Personally, I'd settle for the fountain of middle age, but for those who seek immortality there are two options.

The first is to live long enough to be revivified by up-and-coming technology. Ray Kurzweil, scientist, inventor, and futurist holds that the requisite paraphernalia will be available before 2050. Kurzweil received, among other awards, the 1998 "Inventor of the Year" award from MIT, and he's currently Google's chief engineer, so maybe he knows what he's talking about. On the other hand, he's been known to ingest hundreds of supplements and ten cups of green tea a day in an effort to reprogram his biochemistry[36], so perhaps not.

The second option is to masquerade as a *Turritopsis dohrnii*, the immortal jellyfish (not making this up). *Turritopsis dohrnii* can, after reaching maturity, slowly revert back to its youth and begin the entire cycle again, a process that in theory can go on forever, though most of the jellyfish succumb to predation or disease.[37]

---

### *A Career Path for Jellyfish*

I suggest there's another viable cause for jellyfish demise, one that relates to the philosophy of eternal life, to the story of moi, and to a possible career path for jellyfish. Simply put, I think *Turritopsis dohrnii* dies of boredom. As journalist Herb Caen has pointed out, "The only thing wrong with immortality is that it tends to go on forever."

Consider the life of jellyfish. First, they're not fish; they're not even vertebrates; they're spineless. Second, their nervous system is comprised solely of nerves that stretch along the outer body, with no brain; they're brainless. Third, nobody likes jellyfish; they're unpopular. And finally, they're gelatinous blobs that float around; they're do-nothings. That's why I think jellyfish would thrive in the United States Congress—as spineless, brainless, unpopular, do-nothing gelatinous blobs, they'd fit right in. And whereas a typical Congressperson, being human (more or less), can serve for perhaps fifty years at most, an immortal jellyfish could serve forever, thus sparing constituents the hassle of thinking about whom to vote for. And for the jellyfish, though saddled with immortality and Beltway traffic, they at least would have good perks, like interns and free airport parking. (If comparing Congress to jellyfish seems disrespectful, I'd like to apologize to the jellyfish.)

Many scientists believe the universe will die, and most blame it on thermodynamics. As the universe expands, approaching maximum entropy, all remaining order will dissipate; galaxies and stars as well as planets and their inhabitants will be long gone. That's the most likely endgame scenario, and it's called the Big Freeze. The Big Rip posits that dark energy, the mysterious force accelerating universal expansion, will shred the universe down to nothing more than unbound particles and radiation. There are cosmologists (who aren't such Debbie Downers) that say the universe, like a cosmic *Turritopsis dohrnii*, will cease its expansion and start shrinking, to begin its life cycle again: a Big Crunch will contract the universe to its original state, to be followed by another Big Bang. Scientists call the cyclical succession of Big Crunches and Big Bangs the Big Bounce, but since in not one of the above scenarios does humanity survive, I say Big Deal.

Every living thing is killed indirectly by entropy, but the explicit agent of our death resides within. If a tortoise can live past one-fifty, and a bristlecone pine five thousand years, why do we get shortchanged? Where's the rulebook? What hand dials down the power?

It's our genes, ye of short attention span! The rules are etched in our DNA. Genes assemble us, literally, from the ground up (because we're engineered using terrestrial molecules made organic by plants). Genes build *us* to pass *them* along, and as long as we're having sex they (not to mention we) are happy. Genes, like the Vatican, do not favor contraception. We're their survival machines and they survive when we reproduce, which is why genes, without intention and in a mere nine months, constructed each of us.

Old man: "I'm near death, but I'll live on in my children." No, you won't. You'll be dead as a doornail, 100% atomized, mind, body, and soul, but exactly half your genes will adventure on. Genes hope your children too will dig sex (a pretty safe bet) so that they (the genes) can sashay on down the line. From Dawkins's perspective, genes are the true immortals. Consider, for example, that a) humans and chimps share well over 90% of their genomes, and b) humans and chimps split off from a common ancestor at least five million years ago. Many of our genes date back further than that. We're temporary transport units, discarded after our pay-it-forward purpose has been served.

Like Dawkins, those who've followed up on his reasoning often take pains to emphasize that genes are not selfish, they have no hopes, they take no position on birth control. Genes operate via a completely mechanistic process, a biochemical algorithm; they're the impersonal implements of natural selection, whose basic imperative is publish or perish. At the macro level, genes thrive when they build a superior survival vehicle for themselves, which would be you and especially moi. At the micro level, the double-helix chain of genes we call DNA is nothing but a recipe that cellular chemicals mindlessly follow in order to assemble and maintain their ride—us. Inscribed somewhere in the DNA instruction set is this: When the fools stop having sex, off 'em. If you seek longevity, the lesson should be clear.

# EVOLUTION

*"The universe we observe has precisely the properties we would expect if there is, at bottom, no design, no purpose, no evil, no good, nothing but blind pitiless indifference."*
- Charles Darwin

*"When its whole significance dawns on you, your heart sinks into a heap of sand."*
- George Bernard Shaw, playwright (on Darwinism)

*"If evolution really works, how come mothers only have two hands?"*
- Milton Berle, comedian

---

## NATURAL SELECTION

We've all seen the poster. Not the one of Albert Einstein sticking out his tongue, and not the one of John and Yoko naked, and definitely not the movie poster for *Twilight* (slogan: Vampires Live Forever, which is how you'll feel watching this film). I'm talking about the evolution poster where a stooped-over simian, in a succession of five or six images, straightens up and morphs into a human. There are variations in which the end result turns out to be a golfer, a rock guitarist, a couch potato, and even Kate Beckinsale (in a pitch for *Underworld: Evolution*).

With the exception of Kate Beckinsale, not one of the images in any of the posters represents evolutionary "progress," as persistently pointed out by Stephen Jay Gould and other scientists, yet somehow the heedless mechanism we call natural selection has led us to believe that humankind is its pinnacle. Pinnacle of what, happenstance? But for an exceptionally bad break dinosaurs would be lording it over us, if there had even turned out to be an us. As Bertrand Russell observed, "Since evolution became fashionable, the glorification of Man has taken a new form."

Is it a fact that the Earth is round? If so, then evolution too is a fact. Evidence cascades in from geography, anatomy, geology, fossils, embryology, anthropology, chemistry, carbon clocks, molecular clocks, radiation clocks, genome sequencing, and more. Natural selection is a theory, Darwin's theory, to account for evolution. Darwin envisioned nature as a competition for resources that favors individuals best suited to their environment, and in which, via reproduction and inheritance, adaptive variations proliferate at the expense of less advantageous traits.

The notion of natural selection has now been extended beyond biological evolution, because it's commonsense that within any resource-constrained context, where entities propagate in differential number based on expedient attributes, some will preponderate while others fade. What else could happen? A conga line?

The amplified version, which scientists call Universal Darwinism, impacts psychology, medicine, cosmology, software design, nutrition, and several other fields. I'm probably best qualified to discuss genetic software algorithms, but I'd rather talk about the caveman diet because (and I speak from experience) it makes for livelier cocktail party banter.

The caveman or "paleo" diet is based on the presumption that evolution prepared our digestive systems for hunter-gatherer foods, such as wild plants and meat, but not for agricultural products such as wheat, legumes, and dairy, which have been around less than 12,000 years. If you're considering the paleo diet, here are a couple of genuine caveats. First, the caveman diet is great if you plan to live to 35, the average lifespan of cavemen and Neanderthals. They didn't survive long enough to develop the diseases associated with profuse consumption of animal protein and fat, and in truth, human life expectancy began to rise only after the dawn of agriculture. Second, actual cavepersons consumed up to five or six pounds of greens a day. If you try that, you'll know what a day is.

George Bernard Shaw may be right about Darwinism: "There is a hideous fatalism about it, a ghastly and damnable reduction of beauty and intelligence, of strength and purpose, of honour and aspiration, to such casually picturesque changes as an avalanche may make in a mountain landscape."[38] Personally, I have no issue with casually picturesque change or accidental beauty; in fact I'm not sure there's any other kind. Does it have to be about us?

Religious fundamentalists may be correct too, to deny Darwin and to fight him in the schools. Darwinism is an ideologic universal acid that, "eats through just about every traditional concept, and leaves in its wake a revolutionized worldview, with most of the old landscape still recognizable, but transformed in fundamental ways."[39] Ironically, where creationists might recognize Darwin's overwhelming implications, many who believe in evolution do not. As Shaw warned us, "It seems simple, because you do not at first realize all that it involves." He was at least implicitly on the mark, in that, "full acceptance of Darwin's insights will necessitate revisions in the classical view of personhood, individuality, self, meaning, human significance, and soul."[40]

Is there an escape hatch? Do we have some wiggle room? Is there some way to face up to Shaw's incidental beauty and Darwin's blind indifferent universe with our dignity intact? My answer is yes, with stipulations—groovy stipulations. You could skip ahead to the book's final chapter and check it out, but let me suggest that you don't. If you leap that far ahead the thread of our story will be broken; you'll bypass key milestones and important signposts will go unread. It won't make sense. Plus, you'd miss our next topic.

## WHAT IS SEXY?

---

### Two Kinds of Sex

Every organism exists thanks to reproduction, of which there are two kinds: asexual and sexual. Many bacteria, some plants, and a few insects can clone themselves without a partner, but most humans prefer sex. One possible exception comes up in a snippet of dialogue from the 1983 movie *WarGames*, in a scene between a high school student played by Matthew Broderick and his biology teacher.

*(Actual Dialogue)*

Teacher:    Asexual reproduction. Could somebody tell
            me please, who first suggested the concept of
            reproduction without sex?
Broderick:  Your wife?

---

Why sex at all? Why would natural selection favor such a messy and dangerous, not to mention psychologically bedeviling technique? To make a long story short, and far less titillating than one might have hoped, combining two sets of DNA has survival advantages. It increases genetic variation, promotes DNA repair, and helps mask gene mutations.

To philosopher Daniel Dennett, some mechanisms are "skyhooks" and others "cranes." A skyhook is a contrivance, physical or ideological, that dangles unsupported from the heavens, miraculously able to upraise an agenda and its followers. Of course, lacking support of any kind, a skyhook—call it Elvis if you wish—is an impossibility. A crane, on the other hand, is an apparatus that rests firmly on evolutionary ground, a fortuitously adaptive subprocess that can speed up the protracted course of natural selection. Dennett calls sex a crane. Species that reproduce via sex evolve more efficiently than species that reproduce asexually.

Darwin proposed natural selection in his most famous book, *On the Origin of Species*, but he wrote other books, including *The Expression of the Emotions in Man and Animals*, wherein he described the evolution of emotion, and a 900-page tome, *The Descent of Man and Selection in Relation to Sex*. In *Descent*, Darwin described a second trait that (along with environmental adaptation) nature would favor. He called it sexual selection, but we refer to it as sex appeal. Said Darwin: "The sight of a feather in a peacock's tail, whenever I gaze at it, makes me sick!"[41] Why? Because an ornate tail serves no adaptive purpose, which flouts natural selection, and indeed the main predator of peacocks, the tiger, is adept at pulling them down by their tails. But for reasons that are still poorly understood, peahens (female peacocks) are excited by males with large fancy tails, and favor them as sex partners.[42]

Human beings would never be enticed by an impressive tail, but we do primp for sexiness. Psychologists distinguish between honest and dishonest displays. When a woman wears a tight fitting belt around her waist, that's an honest signal of youth, fertility, and the fact that she's not pregnant, but when a man wears a sports coat with shoulder pads, that's a dishonest display of latent strength.

What do women find attractive in a man? (Well it's not cologne, for reasons described in the CNN story below). Studies have shown that ovulating heterosexual women and homosexual men prefer faces with prominent and broad cheekbones, a longer lower face, developed brows, and chiseled jawlines, all of which are testosterone signals.[43] Women, ovulating or not, are attracted to symmetrical faces, an honest display of health. With respect to men's bodies, women, in general, are drawn to a low waist-to-chest ratio (a V-shape), height, erect posture, and less body hair. That and a big penis, as every guy's spam folder makes clear.

Men, overall, are most attracted to women with symmetric faces (same health indicator), full lips (estrogen signal), clear skin, and clear eyes (health again). A University of Toronto study found that men, like women, respond to facial proportion. Body-wise, men are evolutionarily enticed by youth, breast symmetry but not necessarily breast size, the buttocks (an honest display of fat reserves—no joke), a low waist-to-hip ratio, long hair, and a woman smaller than themselves [fill in your own scientific explanation here].

Could all this be cultural? Not really. Similar sexual preferences have been found, with minor variations, across cultures. (Which is not to say that cultures don't differ in their *ideals* of beauty.)

---

### Laws of Attraction

In 2009, CNN aired a story called "The Laws of Sexual Attraction," in which sex therapist Dr. Laura Berman stated that, "We are innately all puppies in heat. There's a whole realm of unconscious scents that we're not even aware we're smelling." She tells us that women can smell a man's testosterone level, that men can smell when a woman is fertile, and that there's a difference between love and raw chemistry (ya think?). In one study Berman had women smell men's T-shirts, and found they were most attracted to the shirts of men with a different major histocompatability complex (MHC) than themselves. (MHC is a collection of genes related to the immune system.) According to Berman, "We unconsciously want to mate with someone who has a different immune system than ours, because that helps with the survival of our offspring." And what could be steamier than that?

Pop Quiz for Women: Prior to sex, which manly attributes do you take into consideration (check all that apply)? ☐His jawline ☐His torso ☐His body hair ☐His penis ☐His major histocompatability complex.

Pop Quiz for Men: Prior to sex, which womanly attributes do you take into consideration (check all that apply)? ☐She's breathing.

Science has revealed that men and women are different. Whereas women need a reason to have sex, men just need a place.

---

## MYTHS AND CONTROVERSIES

While Darwin could observe the resemblance between parents and offspring, he was unaware of the underlying biological mechanisms and he never methodically characterized the similarities. It was Darwin's contemporary Gregor Mendel, the pea-breeding monk, whose painstaking experimentation laid out inheritance patterns in mathematical detail. Darwin felt that parental traits blended like paint in their offspring, but that notion contradicted his own theory wherein mutations afforded the variation required for natural selection. If traits intermingled, mutations would be blended out over time.

Mendel's experiments with over 30,000 pea plants showed that traits were discreet, and were passed on whole between generations. He mused that tiny "elements" carried the information. Mendel's findings, which presaged genes, would have shored up the "blending" inaccuracy in Darwin's theory, but Darwin never became aware of Mendel's work. It wasn't until 1930 that scientists connected Darwin's natural selection with Mendelian inheritance (and the mutation theory of Hugo De Vries), resulting in the "modern evolutionary synthesis."

To this day, misconceptions linger regarding evolution.

| MYTH | FACT |
|---|---|
| It's survival of the species. | Evolution works on individuals via genes. A species can become extinct via competition between individuals. |
| Humans descended from apes. | A species cannot descend from a contemporary. Human beings and apes branched off from a common ancestor. |
| Evolution "progresses." | It doesn't progress from inferior to superior organisms, or toward greater complexity. |
| Natural selection is the only mechanism of evolution. | Genetic drift is another among several. When survival pressures are weak, random changes can carry forward. |
| Evolution results in perfect environmental adaptation. | No. Just good enough to out-perform rivals. ("I don't have to outrun the bear; I just have to outrun you.") |
| Evolution can't explain traits such as homosexuality. | Homosexuality is common in many species. It has been shown to have indirect adaptive value. |

### The Indispensable Pissing Contest

In his 2007 book *The Stuff of Thought: Language as a Window to Human Nature*, Harvard psychologist Steven Pinker gave credit to "the wordsmiths who thought up the indispensable pissing contest."[44] Originally, a pissing contest meant exactly what the term implies, and while it's typically been men who compete, historians have documented female bouts as well. A comic song from 17th-century Belgium describes a match between three women who are trying to impress a man, but the point of the song is ambiguous. Do the women think that Belgian men are easily impressed, or that they're hard to impress? Do they regard the men as avid sports fans, or as hydro-engineering buffs?

Pissing contests have been depicted in classic literature and in masterly paintings, but Pinker, I believe, meant to acknowledge the term's current metaphorical connotation, as when President Dwight Eisenhower refused to confront Senator Joseph McCarthy because, "I just won't get into a pissing contest with that skunk."

During the final quarter of the twentieth century, evolutionists Richard Dawkins and Stephen Jay Gould engaged in one of modern science's most public pissing contests. (Kim Sterelny's book *Dawkins vs. Gould: Survival of the Fittest* was an international bestseller.) Dawkins posited that evolution proceeds gradually and doggedly forward, based solely on random gene mutations. The Gould camp called this portrayal "evolution by creeps," espousing instead punctuated equilibrium, in which periods of rapid evolutionary change alternate with longer periods of relative evolutionary stability. The Dawkins crowd referred to Gould's construal as "evolution by jerks." Dawkins is a gene's-eye view guy, while Gould put more emphasis on individuals, but neither apple fell far from Papa Darwin's tree, and some regard their apparently divergent views as a false dichotomy.

### DNA

Charles Darwin's grandfather Erasmus Darwin conjectured in 1794 that, "One of the same kind of living filament is and has been the cause of all organic life."[45] Not only did Grandpa Darwin surmise that all life had a common undergirding, he anticipated its shape. Surprisingly, his grandson Charles failed to recognize the significance of this insight until several years after his birth.

The handbook of life is written in a code called Deoxyribonucleic acid (DNA), and with the exception of RNA viruses, every living thing is built and sustained when cellular chemicals perfunctorily follow the handbook's instructions. DNA is a meters-long molecule shaped like a long twisted ladder, and it can be found coiled up within the nucleus of every cell in our bodies. If we were to imagine its "rungs" to be the size of rungs on a household ladder, the magnified DNA molecule would reach to the moon.[46] In 1953, James Watson and Francis Crick proposed the first accurate double-helix model of DNA structure, for which they received a Nobel prize in 1962.

A filament of DNA is built up from four molecules only, in a code that consists of the letters A, C, G, and T: adenine, cytosine, guanine, and thymine, nitrogenous bases that form interlocking pairs and stack together helically. Each base is a molecular match for only one of the others. An A molecule and a T molecule fit together, as do a C and G pair. Bases on each side of the ladder match up with their partners on the other side, forming the two "complementary" strands that comprise a filament of DNA. When the strands split apart, a new complementary strand can form on each separated fibril, and we end up with two copies of the original DNA filament. Replication of a DNA molecule represents core reproduction, its consequence inheritance, while the fact that DNA copies are not without occasional error is the crux of evolution.

Our DNA replicates every time a cell divides, which occurs most frequently during gestation. Each of us starts out as a solitary cell, called a zygote, that's formed via the union of an ovum cell and a sperm cell. The zygote contains our combined parental DNA, which immediately initiates the process of cellular division and differentiation that over a nine-month period ends up in the likes of you and me.

---

### Where Children Really Come From

DNA replication, zygotes, and gestational differentiation are just a theory, the exact sort of prurient gobbledygook that kids encounter in sex education class. Normal adolescents, but for having words like "sperm" crammed down their throats, would show little interest in sex. I'm with those who say fairness mandates we teach a second option—that children come into the world by the grace of an agent from on high. The stork. Fact #1: Unlike some entities, storks have been shown, scientifically, to exist. Fact #2: If sexual intercourse accounts for babies, why do we see so many cases where coitus doesn't lead to childbirth? Fact #3: Data from the Netherlands confirm that as the number of storks in Holland has declined, so has the number of newborns.[47] Unconvinced by the stark stork facts? Then turn skyward for an answer, and the heavens will give unto you the bird.

---

## FABRICATING EVOLUTION

The purposeless genetic flux we call natural selection has been the primary mechanism of evolution, but along the way people figured out how to actively manipulate the gene pool. Darwin wrote a book about "artificial selection," a process we now call selective breeding, and it was

Mendel's meticulous cross-fertilization of pea plants that led him to his principles of inheritance. But human beings had domesticated animals and plants long before Darwin and Mendel.

Recent DNA evidence suggests that dogs descended from a wolf-like ancestor that became extinct 10 or 20 thousand years ago. One theory holds that dogs hung around human encampments to scrounge for refuse and scraps, and that those who were comfortable near people had greater success, were healthier, and bred. The genes that inclined them to tameness were disproportionately passed on, then subsequently cultivated through breeding. Paleo-Fido provided early humans with a guard animal, a source of food and fur, and a beast of burden.

Cats sort of domesticated themselves. Why? They can't open cans. About twelve thousand years ago, at the dawn of agriculture, the ancestors of today's tabbies wandered into Near Eastern settlements looking for food. What they found were rodents, lots of rodents, feeding on stored grain. The cats who could tolerate human contact caught more vermin, survived in greater numbers, and passed on their genes. At some point cats were docile enough for breeding. From the feline point of view humans had opened up a rodent food court, and from the human perspective, Proto-Fluffy was the original granary guard.

Today's scientists can bypass breeding and directly manipulate DNA, a process known as genetic engineering. Once foreign or synthetic DNA segments have been introduced into an organism's natural DNA, replication will pass the modified figuration along. Genetic engineering, as you know, is a divisive issue.

More controversial, however, is eugenics, the purposeful genetic "improvement" of a human population. During the early 1900s, eugenics was practiced in the U.S. and around the world via marriage restrictions, segregation, compulsory sterilization, mandatory abortion, birth control, forced pregnancy, and genocide. Its popularity waned after World War II, largely due to Nazi Germany's barbaric efforts to cull a master race, though Sweden maintained a eugenics program until 1975.

Dysgenics, originally viewed as the antithesis of eugenics, refers to a population's unintentional genetic decline. Sir Francis Galton, who coined the term eugenics, was the first to broach the now-incendiary subject of dysgenics. In 1869 Galton wrote, "There is a steady check in an old civilisation upon the fertility of the abler classes: the improvident and unambitious are those who chiefly keep up the breed. So the race gradually deteriorates, becoming in each successive generation less fit for a high civilization."[48] Today, there's a fierce scientific and moral debate as to whether human IQ will decline as the less intelligent outbreed the intelligentsia.

---

### Galton Stands with Beavis and Butthead

The 2006 movie *Idiocracy* contends that IQ will in fact plummet. This is the film's opening voiceover narration:

> As the 21st century began, human evolution was at a turning point. Natural selection, the process by which the strongest, the smartest, the fastest, reproduced in greater numbers than the rest, a process which had once favored the noblest traits of man, now began to favor different traits. Most science fiction of the day predicted a future that was more civilized and more intelligent. But as time went on, things seemed to be heading in the opposite direction. A dumbing down. How did this happen? Evolution does not necessarily reward intelligence. With no natural predators to thin the herd, it began to simply reward those who reproduced the most, and left the intelligent to become an endangered species.

The movie was written and directed by Mike Judge, who created Beavis and Butthead, so I don't have to tell you that the film is 99% potty-mouth farce and 1% scientific mumbo jumbo. The plot involves Private Joe Bauers (Luke Wilson), who awakens 500 years in the future after a botched hibernation experiment. He's accompanied by a hooker named Rita (Maya Rudolph), who volunteered for the experiment to avoid prosecution. On waking, Joe discovers that he's by far the smartest person in the United States, which via dysgenic degeneration has become a nation of morons.

The film fulfills Galton's longstanding prophecy, but I'm not sure Sir Francis would see the humor in imbeciles. Americans apparently do, because *Idiocracy* garnered a 73% on the Rottentomatoes tomatometer, which is why I opted to watch the movie. Bad idea. To me, it's evidence that dysgenic deterioration has already set in.

Surgeon General's Warning: If you suffer from acute political correctnessitis, do not read the remainder of this section unless you have an airplane sickness bag within arm's reach.

Have you wondered why the CDC, in 2011, issued *Preparedness 101: Zombie Apocalypse* (which I'm not making up)? Well it wasn't to prepare us for zombies, which would be ludicrous, but it's as close to the truth as political correctness allowed. The CDC is in fact prepping us for the already-in-progress Dummy Apocalypse—look around!—it's *Night of the Living Dumb, DummyLand, World War D.*

It's clear that government has been infected, one more reason for the CDC's transparent "zombie" ruse (transparent to everyone except dummies). Are you aware that there have been more than 1,800 books published specifically for dummies, including *Dating for Dummies, Ballet for Dummies, Football for Dummies, Feng Shui for Dummies,* and *Menopause for Dummies?*[49] How dumb will Americans be in another five hundred years? To hear Mike Judge tell it, these are the two smart ones:

*(Actual Dialogue)*

Rita:    You think Einstein walked around thinkin' everyone was a bunch of dumb shits?
Joe:     Yeah. Hadn't thought of that.
Rita:    Now you know why he built that bomb.

Plus, you know why the Surgeon General issued a warning, and why cinematic evidence says dysgenic decline has already kicked in.

## BYPRODUCTS OF EVOLUTION

According to strict adaptationists, evolutionary attributes result solely from selective environmental accommodation. Traits evolve to meet an ecological need. In an influential 1979 paper, Stephen Jay Gould and Richard Lewontin argued that some traits develop indirectly, as side effects of primary features. They used the architectural term "spandrel" to refer to a characteristic that's a byproduct of the evolution of some other trait. They asked, for example, why blood is red. Is that pigment advantageous, or is it merely a consequence of the fact that hemoglobin happens to be red?

Gould, in 1997, offered a somewhat more piquant example that highlights the sexual predicament of a female spotted hyena. Among hyenas, females are larger, more muscular, and dominant, because fetal females are pumped up with androgen, a male sex hormone designed for aggression. They literally come out fighting. Androgen is adaptive since female hyenas compete ferociously for food and to protect their young, but there's a side effect that can be summed up in the words of Michigan State professor Kay Holekamp. "Imagine giving birth through a penis."

The clitoris of a female hyena protrudes seven inches from her body, and is almost indistinguishable from the penis of a male. Her birth canal is an inch in diameter, and the sex act itself is beyond awkward. According to Holekamp, "Males need practice. After a couple of months of practicing, they get it lined up just right."[50] [Fill in a humorous analog of your own here.] Gould and Holekamp insist that a seven-inch long, inch-wide clitoris is an evolutionary spandrel, a byproduct of androgen's survival utility.

My girlfriend and I have four cats. I love them, but why? Cats are merciless predators, both in our human past and in the wild today. Of the four most dangerous man-eaters three are cats—lions, tigers, and leopards (crocodiles are the fourth).

To me, love of cats, or of any nonhuman species, is a spandrel. I don't know how to explain it unless my innate bonding mechanism just happens to spill over, as a fortuitous evolutionary side effect, onto cats. Why else would I rather watch our cats than watch the Dallas Cowboy Cheerleaders? (Granted, a Dallas Cowboy Cheerleader has never curled up in my lap or licked my hand, though I'm open to the idea.) Does it really seem normal for a nuclear family to include multiple species?

According to the Pew Research Center, 85% of all dog owners consider their pet to be a member of the family. Cat owners chime in at 78%. Ninety-four percent of dog owners describe the relationship with their pet as "close," as do 84% of cat owners. When the same owners were asked to describe the relationship with their parents, 87% described Mom as "close," but Dad trailed the entire mixed-species menagerie at 74%.[51]

## BEYOND EVOLUTION

Have we reached the end of natural selection? Has evolution fashioned a mutinous creature, able to overthrow its creator? Can we, the vehicles, wrest control of our own steering wheels?

Technology might be the key. Perhaps humankind can overcome evolution and even death via genetic engineering, nanotechnology, stem cells, AI, informatics, and other brainy new technologies. Soon, bionic implants, organ transplants, and mind-to-computer uploads could keep us, or some version of us, going indefinitely. That's exactly what Ray Kurzweil, Google's chief engineer, forecasts. He predicts that by the year 2050 we'll be able to upload our minds to a computer, and that by 2100 our bodies will be replaced by machines. Right. That's how I want to end up—a rusty, hard-shelled, fuse-blowing automaton. I know exactly what my friends will say when they see me. "Hey Hick, haven't changed a bit."

There's another way humanity can cast off evolution: extinction. Scientists estimate that 99.9% of all species that have ever existed are now extinct. At this moment there are so many species vanishing, an estimated three per hour, that some scientists believe we're at the onset of Earth's sixth mass extinction. The most recent event, known as K-T, wiped out the non-avian dinosaurs 65 million years ago, tidying up the food chain for the ascendance of mammals. Paleontologist Doug Erwin, a specialist in mass extinction, suggests that though such an event might be at hand, the evidence so far isn't conclusive. If a mass extinction is indeed underway, Erwin recommends "going out and buying a case of really good Scotch, because we're screwed."[52] Cosmologists would tell us that extinction is guaranteed, since entropy will ultimately dissolve our universe. Cosmetologists, no doubt, would suggest we look good for the occasion.

What actually killed the dinosaurs? Their demise, it turns out, has once again become a bone of scientific contention. Until recently it appeared that science had settled on the Chicxulub theory, Chicxulub being an impact crater on the Yucatan Peninsula and not a gangrenous salsa or Mayan curse. Under that scenario, the asteroid's ten-mile-wide concussion would have generated firestorms across the continent and coughed up a planet-enshrouding, climate-altering dust cloud. Or not.

We could look to Mark Graham, musician and comic songwriter, whose album *Natural Selections* features numbers like "The Big Band Theory" and "I'm Working on the Food Chain." In his dinosaur song, Graham tunefully ponders what we actually know about the behemoths, and to go by the song's title our dinosaurian acumen extends to "Their Brains Were Small and They Died."

Some of us, though, enjoy a more Flintstones-flavored sensibility. A National Science Foundation study found that 49% of adult Americans believe humans and dinosaurs coexisted.[53] Former Vice Presidential candidate Sarah Palin declared that, "dinosaurs and humans walked the Earth at the same time," and that she had personally "seen pictures of human footprints inside their tracks."[54] The 70,000 square foot Creation Museum boasts a diorama in which, "children play and dinosaurs roam near Eden's Rivers."[55]

Fast-forward to modern times, in fact to our not-so-distant future, which is where Graham's dinosaur song ends. Two cockroach scientists peer out over the ruins of human civilization.  Explains one:  "Their brains were small and they died."

# THE BRAIN

*"Almost every aspect of transcendent human experience, including love, memory, dreams, and even our predisposition for religious thought, ultimately derives from the inefficient and bizarre brain engineered by evolutionary history."*
*- David J. Linden, neuroscientist*

*"What each of us regards as his or her own intimate private self is simply the activity of these little specks of jelly in our heads, in our brains. There is nothing else."*
*- V.S. Ramachandran, neuroscientist*

*"Brain: an apparatus with which we think we think."*
*- Ambrose Bierce, author*

---

## BRAIN IN A VAT

### Dude, Where's My Brain?

How many angels can dance on the head of a pin? What's the sound of one hand clapping? Philosophers ask tough questions, and their answers tend to be simultaneously erudite and enigmatic. In this case the correct answer to both questions happens to be the same: All. Early in the book I said there'd be no magic, liturgy, or new age prattle, but I fear we must now teeter out onto the high wire of philosophical reckoning. We need to confront the Evil Genius.

Try not to stare at the accompanying diagram. True, in its unadorned form it should induce only mild anxiety, but were we to add a caption, such as, "plot a graph of the parabolic function $f(x) = 2x^2 + 3x - 5$," nauseating flashbacks might surely set in.

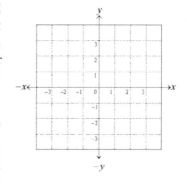

Fortunately, in our era Americans have figured out how to circumvent this hazard. Thanks to mass communication and cultural diversity, we now know that others in our world can actually stomach this sort of drudge. Some seem to thrive on it, which is why the United States, in what could only be described as a win-win coup, has successfully outsourced mathematics.

The image, as you can reluctantly recall, depicts the Cartesian plane, with the hash marks representing Cartesian coordinates, but Descartes, in spite of his never-ending torture of high school sophomores, is not the Evil Genius. It was Descartes, however, who invented the Evil Genius. One of the cornerstones of Descartes's philosophy is an approach we now call Cartesian skepticism. Basically, Descartes tried to discover what was authentically true by systematically doubting all his beliefs, hoping to cull the ones that could withstand unremitting distrust. The famous pronouncement, "I think, therefore I am," (1637) is one of his "irrefutable" truths. Descartes tried to doubt his own existence, but found that even his doubting showed that he existed, since he couldn't doubt if he did not exist. In 1846, Søren Kierkegaard pointed out the tautological nature of this reasoning (italics his). "If the *I* in *cogito* is understood to be an individual human being, then the statement demonstrates nothing: I *am* thinking ergo I am, but if I *am* thinking, no wonder, then, that I am; after all, it has already been said."[56] (See what I mean? Erudite yet elusive.)

Descartes's "Evil Genius" is a thought experiment designed to challenge what is real. He posited an Evil Genius who was supremely clever, powerful, and deceitful, and whose sole and consuming effort was to mislead Descartes. To that end, the Evil Genius constructed around and within Descartes, at all times, a comprehensive illusion of the world, complete with thoughts, emotions, other people, their thoughts, the illusory sensation of a physical body, the works. Descartes asked us then if there could be any way to tell the difference between a reality so induced and a genuine world.

Descartes paid a price for his grandiloquence. Protestants, at least the ones who could understand what he was talking about, accused him of blasphemy for conceiving an omnipotent God of malevolent intent.[57] I think they missed the point.

This might be a good time to introduce a useful technique for deciphering esoteric terminology, which is to split the term into its component parts. The word grandiloquence, for example, can be broken down into: *grand / i / loquence*. Grand, of course, means big or overblown, I refers to me, and "loquence" connotes flowery speech, as in the word eloquence. By that analysis, what Descartes had to pay for was his grandiose, self-centered rhetoric. [Philosophy buffs are welcome to insert a 30-page rebuttal here.]

All of which brings us to "brain in a vat," the modernized version of Descartes's Evil Genius. Here, philosophers imagine a brain suspended in a vat of life-sustaining fluid, with electrodes running from its neural junctures to a supercomputer whose job it is to provide precisely the electrical impulses the brain would receive in a real world. Under those circumstances, they ask, how could a brain know whether it's in a skull or a vat?

And that brings us to *The Matrix*. You know, the movie where Keanu Reeves, upon discovering his whole life to be nothing but a simulation, is so upset that: a) his facial muscles become paralyzed for the rest of his acting career, and b) he can't stop vaulting around in slow motion, shooting at everything in sight. That second part reminds me of something a cowboy friend of mine once told me (true story) about cowboys. "If we can't eat it, drink it, or screw it, what we want to do is shoot it full of holes." In my buddy's case that would conveniently leave nothing to shoot at but stationary targets, since he'll pretty much eat, drink, or screw anything that moves.

*(Actual Dialogue)*

Morpheus:      Have you ever had a dream, Neo, that you were so sure was real? What if you were unable to wake from that dream? How would you know the difference between the dream world and the real world? You've been living in a dream world, Neo. Welcome to the real world.

And that, dear Neo, brings us back to the story of moi. To this point we've reviewed data which show that each of us, physically and biologically, is the product of unwitting algorithms along with random chance. And most people seem able to accept their physical selves as the accidental byproduct of stellar and biological evolution. We appreciate the material evidence, and we recognize that other accounts necessarily fall back on magic, religion, or simple narcissism.

From a philosophical perspective though, and in my personal observation, the main reason we're willing to write off our bodies to naturalistic happenstance is that most of us are, at heart, Cartesian dualists. (Same guy—he lived in Paris and Amsterdam—couldn't he find something better to do than sit around and think?) It's philosophy's longstanding mind-body problem. Most of us don't feel like we *are* a body, we feel like we *have* a body. We experience ourselves as entities that inhabit our bodies.

To locate precisely where your inner self abides you can try the following exercise. It's for real, it's called finding your "point zero," and you don't even have to get up off the couch to do it. Put your right hand in front of your forehead, and with your index finger point to the place inside your head where it feels like "you" are right now, then with your left hand to the side of your head, point again to where it feels like "you" are. The intersection of the two lines is point zero, where your "self" hangs out, and with that you'll have answered one of life's great questions. Where's Waldo?

At this point in our story it's going to get personal. We'll get over ourselves later, but for now we'll be talking directly about the indwelling self. Actually, that's not true about this particular chapter, which is the book's transition from physical and biological science to psychology. In the next chapter we'll segue from the brain to the mind, then to emotions, consciousness, and selfhood, the inner experiences that most of us relate to as ourselves. Ultimately, it's over-identification with exactly these subjective realities that we'll need to step back from, and for the remainder of our story science will help illuminate why and how to do that. In our final chapter we'll roll out "A New Moi."

But just as in *The Matrix*, we may not like everything we see. For a moment let's pretend I'm Morpheus (Laurence Fishburne). You get to be Neo (Keanu Reeves) or Trinity (Carrie-Anne Moss), and you should feel free to gender-bend. Or you could opt for the role of Cypher (Joe Pantoliano), who betrays Morpheus, Neo, and Trinity because he wants to get back *into* the Matrix, even though he knows it's a complete fabrication. "I know this steak doesn't exist. I know that when I put it in my mouth, the Matrix is telling my brain that it's juicy and delicious. After nine years, you know what I realize? Ignorance is bliss."[58] We're all tempted to go that route, but we don't need to. The truth will set us free.

Since we're role-playing, you could visualize yourself in a different film altogether. You could be the world's greatest brain surgeon, played, before you nailed the part, by Steve Martin in Carl Reiner's *The Man With Two Brains*. You could, following the death of your first wife, get swept into marriage by a gorgeous gold digger (Kathleen Turner) who just happens to experience a headache whenever it comes to having sex with you. Of one sexual episode you might remark, "That was the most exciting sexual encounter, without actually having it, that I ever almost had."[59] Eventually, you might find yourself drawn to a disembodied but communicative female brain that you'd been carrying around in a bell jar. "Now, for the first time, I'm aroused by a mind." Given the following set of plot points, see if you can figure out how the movie ends: 1) you're married to a comely but wicked wife, 2) you're in love with a bright and kindly brain that lives in a jar, and 3) you're the world's greatest brain surgeon. If you didn't decide to put the compassionate brain in Kathleen Turner's body then you need to buy my other book, *Screenwriting Made Obvious.*

Is any of this scientifically plausible? Should we, as Steve Martin's comically bombastic character put it, "envision a day when the brains of brilliant men and women can be kept alive in the bodies of dumb people"?[60] For now the answer is no. Brain transplant would be far more complex than the sorts of organ transplants we do today, but scientists and doctors are already working in that direction. Dr. Robert White, an actual lifelong brain surgeon, calls brain transplant, "an operation of the future," and he's thought through how it might work.[61] But as you can imagine, piled on top of medical and scientific challenges are social, ethical, legal, and spiritual issues. Some aspects of the matter seem ghoulish, which is why I want to reassure you, as you read on, that mad scientists aren't going to sneak in, remove your brain, and stick it in a vat. We did that three months ago.

## UNINTELLIGENT DESIGN

Answer the following multiple choice question. Look for hints, such as the little numbers that follow the first two options.

The human brain has been described by neuroscientists as:

a.   The most complex organized structure in the universe.[62]
b.   A cobbled-together mess.[63]
c.   Both of the above.

Evolution is a tinkerer, not an architect, and though the human brain can be impressive, it evolved in exactly the same fashion as the rest of biological existence—via ongoing, imperfect, but somehow workable fine-tuning. We'll consider a famous example of "unintelligent design" in detail, briefly present a couple of more colorful illustrations, then turn to our paradoxical brains.

In mammals, the laryngeal nerve has four branches that run from the brain to the larynx. Two of the four branches end up on each side of the larynx, but the two branches that end up on the same side arrive via rather different routes. The left and right superior laryngeal nerves go directly from the brain to the left and right sides of the larynx. The left and right recurrent laryngeal nerves instead exit the brain and veer into the chest, where each recurrent laryngeal nerve then loops around an artery of the heart before doing a 180 and heading north to the larynx. In humans, the detour amounts to perhaps six inches, but a giraffe ends up with a fifteen foot nerve where a one-footer would do.

That's not a problem; the thing still works; it's nothing more than an inefficiency. Unless, that is, you happen to be a surgeon operating in the vicinity of the improbably meandering recurrent laryngeal nerve. Surgeons who perform operations involving the lower cervical spine (the vertebrae that connect the skull to the spine) are specifically cautioned to avoid injuring the recurrent laryngeal nerve.[64]

How did evolution arrive at such an odd adaptation? Scientists tell us it goes back to the common ancestors we share with fish, where the recurrent laryngeal nerve traveled directly from the brain, past the heart, to the gills, as it does in modern fish. You rarely notice a fish wearing an ascot because fish don't have necks, but mammals do. After mammals diverged from early fish and began to evolve independently, their necks extended, their hearts lowered, nerves and blood vessels stretched out, and Macy's introduced the first cravat. At some point in this protracted repositioning, the recurrent laryngeal nerve got caught on the wrong side of the heart.

Natural selection proceeds in tiny increments, and instantaneous replumbing is not an option. Any such gross mutation would surely kill its host. In creatures that survived mammalian evolution, including us, the gradual lengthening of the recurrent laryngeal nerve resulted in the circuitous route we find today.

Charles Darwin, who boasted a bulbous nose, once wrote, "Will you honestly tell me whether you believe that the shape of my nose was ordained and guided by an intelligent cause?"[65] The astrophysicist Neil deGrasse Tyson too has a few favorite examples of dubious evolutionary design.

- We eat and breathe through the same hole. Brilliant! Inadvertent choking is currently the fourth leading cause of unintentional injury death in the United States.

- Pain bio-gauges: A paper cut hurts like crazy, but you usually don't find out about high blood pressure, colon cancer, heart disease, or diabetes until your doctor (or undertaker) gives you the news.

- "And what comedian designer configured the region between our legs—an entertainment complex built around a sewage system?"[66]

Here are three reasons why I can't be faulted for the noticeable defects in my brain: 1) During evolution the brain was never redesigned from the ground up; new systems got lumped on top of old ones, 2) The brain has a limited capacity for turning off automatic control systems, such as the ones for sex and aggression, even in situations where they're counterproductive, and 3) Neurons, the brain's basic processing units, are slow and unreliable, with a limited signaling range.[67]

I blame my defective brain on natural selection, where the perfect is never the enemy of the good enough, or in cases like mine the barely passable. The reason human lives rarely roll along on smooth-spinning wheels is that our charioteer, the brain, is a cobbled-together congress of recent, primitive, and competing motives of which the so-called "self" is hardly aware.

## EVOLUTION MADE ME DO IT

Our species, as you know, is called *Homo sapiens*, where *Homo* represents our genus and *sapiens* our species. Here's a tip: If you want to figure out if a particular creature—let's call it Randy—is of the same species as you, try interbreeding with it. If you two produce offspring then Randy is absolutely of your species. If no offspring ensue despite repeated attempts, then Randy is probably not of your species, though it could simply be a case of looking for love in all the wrong places.

---

### A Fork Along the Evolutionary Highway

Poet Robert Frost wrote of a fork in the road. In his poem he chose the path less traveled, which, he tells us, made all the difference. But how? Was there a reason nobody else took that path, or did his free-thinking move pay off? What awaited him down the way, a quicksand pit or a cheerleaders' convention? Six million years ago, CHLCA, the chimpanzee-human last common ancestor, came to a fork in the road, where chimps headed off down one evolutionary pathway, you and I the other. We were wise to follow Yogi Berra's advice. "When you come to a fork in the road, take it."

The genus *Homo* dates back approximately 2.5 million years, to a hominid called *Homo habilis*, which had a brain about the size of a chimp's today. Fast-forward 2.5 million years. Human beings rule the planet while chimps are just, well, chimps. Question #1: How did that happen? What made all the difference?

A horse colt walks within two hours of birth, but humans take twelve months or so to begin toddling, then remain largely helpless for years. Question #2: How did evolution come to favor years of childhood dependency over nature's nearly universal fast track to self-sufficiency?

The answer to both questions is "encephalization," a term which refers to the excess mass of an animal's brain in proportion to its total body mass. One measure, called the encephalization quotient (EQ), is the ratio of an animal's actual brain mass to the expected mass for that animal. Dogs and cats, for example, each sport an EQ of approximately 1, which means the mass of their brains is what you'd expect in animals their size. The encephalization quotient for chimps and monkeys falls between 2 and 2.5, whales come in at around 3, dolphins between 4 and 5, while humans range from 7 to 7.5.

---

*Baby Talk*

In the movie *Taxi Driver*, Robert De Niro famously asked, "You talking to me? You talking to me?" Queen Latifah posed a related query in 2004's *Taxi*. "You talking to me or your gun? You better not be talking to me." I believe we can assume, despite the fact that infants occasionally pop up (or more accurately pop out) in a taxicab, the actors' remarks were not addressed to newborns.

---

All human babies, whether born early, on time, late, or in a cab, are "premature" for a simple reason: if their brains were any bigger, skull size would make human birth a disadvantageously risky ordeal. That, says biological anthropologist Wenda Travathan, is the reason "having assistance at birth has been favored in our species."[68] In chimps, most brain development occurs prior to birth, whereas natural selection has favored postnatal development in humans, and while a chimp's brain is fully developed within three years, the human brain doesn't mature until we're twenty-five, the exact age at which automobile insurers, perhaps coincidentally, quit charging us premium rates.

No one knows what led to encephalization in humans. Scientists do agree that about 2.5 million years ago an oncoming ice age forced our arboreal forebears down from the dying trees. The selection pressures of ground-based living resulted in major evolutionary adaptations, most notably bipedalism, which afforded our ancestors a better view, freed their hands for foraging and tool-wielding, and required less energy for locomotion. A downside of bipedalism, though, was that it necessitated a narrowing of the birth canal. (Another is that pain in your lower back.) Natural selection's resolution of the small-canal/big-head conundrum turned out to be "premature" childbirth along with protracted human maturation.

Some scientists believe that encephalization was an evolutionary response to the new, complex, and constantly varying circumstances that our anthropoid predecessors encountered while ambling the savannah. Or perhaps it was the development of language and tools that pressured the evolution of brainpower. Some say that intelligence was favored by way of sexual selection, wherein (nerds take note) braininess was taken as a sign of reproductive fitness. And then there's the popular social brain hypothesis, which posits that prehistoric humans, once out of the trees, banded together in larger groups and settled in one place longer. The new and intensifying complexities of social interaction, including rivalries, alliances, pecking order, and courtship, demanded ever more sophisticated interpersonal astuteness. Hence, we grew a social brain. And because childhood dependency ordains that each human self be pieced together in juxtaposition with others, we grew a social self.

And what could be more social than cooking for the clan? *Homo sapiens* is the only species that cooks its food, and primatologist Richard Wrangham avers that fire, in conjunction with the consequent profusion of cooked food, is what propelled humanity forward. Our ancestors spent less time foraging, chewing, and digesting, all of which freed up metabolic resources for brain growth. In Daniel Dennett's terminology cooking would be a "crane," a fortuitous development that helped speed up the prolonged course of natural selection, and indeed, there are only two proficiencies that set humankind apart from other species: language and barbeque.

Encephalization, though a heady episode in the chronicle of you and me, is only our brain's most recent makeover. Paul MacLean, in the 1960s, formulated a "triune" model of the brain that relates cerebral structure to evolutionary advancement. His "three brains in one" theory is where we get the term reptilian brain, which we now know is the part of the brain that controls male behavior. (Though some say the part that controls male behavior isn't even located in the brain.)

Neuroscientist David Linden encourages us to think of the brain as a three-scoop ice cream cone. The bottom scoop, which resides in the brain stem, is called reptilian because it works the same way in a human as it does in a lizard. Evolutionarily, it's the most ancient part of the brain. The reptilian brain regulates vital functions such as breathing, heart rate, blood pressure, and digestion, and since I personally would have difficulty remembering to breathe, it's fortunate that the brain stem performs its rather important work without me being involved.

The next scoop up is the paleomammalian brain or midbrain. The midbrain houses the limbic system, which is responsible for creaturely survival and the emotions that drive it. Our paleomammalian brain is similar to a feline's, and its main responsibilities include the four F's: fighting, fleeing, feeding, and the other primal activity that begins with an F.[69] The midbrain evolved in our early mammalian ancestors, and like the brain stem, it operates largely without the intervention of what we most often regard as ourselves. Having said that, we humans expend considerable conscious energy in both catering to and warding off our animal impulses.

The top scoop, the repository of cognition and awareness, is the neomammalian brain or neocortex. Our cortex is the dual-hemispheric sheath overlaying the brain, folded on itself in layers so that more tissue can be crammed into the skull. The cortex, which is where language, imagination, abstract thought, and consciousness abide, evolved in our primate ancestors then developed disproportionately in human beings via encephalization.

---

### A Party in My Head

The three parts of the brain operate somewhat autonomously, yet they actually interconnect. In the following example of compartmental interconnectivity, I'll use the general triune terms hindbrain, midbrain, and forebrain.

Let's say you're at a party when the visual cortex in your forebrain receives incoming data which suggest the proximity of a potential sex partner. That information is shared with the midbrain—you feel desire. The lust circuits connect back to your forebrain's language centers, which compose a great opening line, but the other rebuffs you for one of three possible reasons: a) The other's midbrain was not activated (feeling not mutual), b) The desire is mutual, but the other's forebrain has reasoned that you're just not worth the trouble, or c) The other's hindbrain has triggered in him or her a vomit reflex. Your own midbrain kicks back in, whereupon you feel angry or hurt, and almost immediately one of several action plans bubbles up to your forebrain: 1) try again/harder, 2) retreat, 3) retort, 4) sour grapes, 5) try that other person over there.

Since we all know the above scenario, perhaps too well, I'd like to review it again from the perspective of self. Initially, someone caught your attention and you reacted. You didn't consciously "decide" he or she was interesting, and you didn't think through how to proceed. Your opening gambit just sort of popped out, and the other's reaction was likewise unrehearsed. You became infused with involuntary emotion and your final action plan spilled forth. The episode is over, an impromptu happening. Your body was an actor in the scene, but your rational self, by and large, was an insentient bystander. Unbeknownst to you, however, your brain recorded the entire incident.

The next day, as you mentally reconstruct the events of the party, you feel as if you'd been put down. You take it personally. Yet, what you experienced the preceding night was nothing more than a natural process unfolding spontaneously. Neither person had a private agenda; your mindful self was barely present. It's over. By the next day the event exists only in your mind. Why react now, and why to a figment of your imagination? Better for all involved if we could realize that this time it's not personal.

## LOOK INTO MY I

During the 1770s, German physician Anton Mesmer used magnets to treat, among other conditions, mental illness. Mesmer believed that a magnetic fluid permeated the entire universe, including our bodies, so he would have patients drink a solution containing iron, after which he'd use magnets to manipulate their internal fluid. Symptoms sometimes improved, but Mesmer at some point decided that it wasn't the magnets after all, but his own "animal magnetism" that had effected the cure, so he doffed the magnets. Instead, Mesmer had patients stare into his eyes as he waved his hands over their bodies. He found that this treatment produced results that were—and no one could have predicted this—just as effective as magnets. The term mesmerism, a forebear of hypnotism, originated in Mesmer's approach, and indeed Mesmer may have been one of the inspirations for the fictional hypnotist Svengali, who seduced and controlled women using only the power of his will. (And I'm not talking about the modern-day practice of promising to make her your heir.)

Magnetic remedies are dubious at best, which is why in our era Amazon's Health & Personal Care department refuses to carry more than 1,320 different magnetic bracelets.[70] In the 1770s, of course, physiological knowledge was imperfect, especially with respect to the nervous system. A few years after Mesmer, Italian scientist Luigi Galvani noticed that an amputated frog leg would twitch when a nerve in the leg was exposed to electricity. Depending on the story you hear, Galvani's ah-hah moment was sparked either by lightning or by an assistant's statically charged scalpel, but in any case he'd discovered an authentically valid principle he called "animal electricity." (Ironically, it's Mesmer's terminology that has endured, although according to the online *Encyclopedia Britannica* the term "animal magnetism" now refers to sex appeal, which means that Mesmer's conception has drifted from being credited for cures, to being discredited by science, to its current level of unrivaled cultural credit.) Fifty years later, following on Galvani's work, physicist Carlo Metteucci measured the bioelectricity by which the nervous system governs our bodies and minds.

The nervous system's control tower is the brain, which comprises approximately a hundred billion neurons along with various support cells. Practically all neurons consist of a cell body that has dendrites on one end (where impulses, which can only travel in one direction, arrive from other cells) and an axon on the other (where impulses propagate forward). In shuttling from one nerve cell to the next, impulses must traverse the tiny interstice, called a synapse, between them.

An abundance of neurons is advantageous, and humans have more than any other animal, but brainpower is not in the neurons per se. Human babies come into the world with virtually their full complement of neurons, yet they rarely give scientific lectures or perform the lead in *Swan Lake*. At birth, a baby's brain weighs only about one-fourth of its adult weight, but during year one the brain makes up more than half that difference, not principally from the construction of new or larger cells, but via the enhancement of neuronal pathways.[71] Networks of connectivity are the key, and neurons build up connections using atoms and molecules, which have mass.

Contemporary research suggests, in fact, that there may be more neurons in a baby's brain than it will retain into adulthood. Through a well-documented mechanism called pruning, underused and inefficient connections are deconstructed. Pruning, which is most evident between birth and puberty, is one of the processes that account for plasticity, meaning that our brains are subject to continuous rewiring, with neural connections strengthening or weakening on a "use it or lose it" basis. The theory, first put forth by Canadian psychologist Donald Hebb in 1949, was later paraphrased by Carla Shatz: "Cells that fire together, wire together."

The bottom line is that with the number of neurons in the brain ("billions and billions"), and the number of possible links (each neuron can connect with up to ten thousand others), the count of prospective network permutations is immeasurable. Some calculations suggest that the number of possible brain states exceeds the number of elementary particles in the known universe.[72]

---

### *Jennifer Lawrence: Brain State or Elementary Particle?*

A brain state is a complex neural network configuration. An elementary particular is a physical unit whose substructure is unknown, such as a quark, one of the building blocks of conventional matter. Until relatively recently atoms were thought to be indivisible, but we know now that atoms are composed of quarks, which constitute protons and neutrons, surrounded by a cloud of electrons. By sharing outer shell electrons, atoms can bond into molecules, that can in turn assemble themselves into temporarily stable configurations. Thus arises the world as we know it, a dominion that includes each of us.

Forms don't create themselves, their molecules self-organize, and though each assemblage exhibits a set of characteristics, they don't *belong* to that form. They're serendipitous attributes of a conferred architecture. For example, I can build a wall thanks to muscle and bone, and I can cogitate courtesy of my brain, but if I can't claim authorship of those bodily mechanisms, how can I declare proprietorship over their artifacts? The wall I build and the thoughts I think are byproducts of a product, second-level fabrications. Wall-building and thought-thinking happen to be capabilities of my bestowed configuration, which is why I find it hard to take credit for them.

No matter how it might seem to me, research shows I have limited control over my sundry brain states. What I consider my "opinion," for instance, exists as a cerebral conformation, potentially subject to neuronal realignment. Whereas today I might think that Jennifer Lawrence is a talented young actress, tomorrow I might contend instead that she's also a knockout and a pip. Brain-wise, we can safely define ourselves as an array of neural patterns that display overall stability, but which are subject to quirky eccentricities.

## ON THE CIRCUIT

The locus of brain activity is the synaptic cleft, the microscopic gap where one neuron's axon terminal abuts the dendrite of another. Spikes of electrical energy, called action potentials or nerve impulses, travel down the axon to the synapse. When the electrical spike reaches the axon terminal, it triggers a series of reactions that result in the release of chemicals known as neurotransmitters into the synaptic cleft. The neurotransmitter molecules migrate across the cleft and bind with receptor sites situated along the cell membrane of the neighboring neuron's dendrite. Proteins in the postsynaptic neuron then convert the chemical information back to electricity, and signals from the dendritic branches funnel toward the cell body. If a sufficient number of pulses arrive at the cell body simultaneously, the postsynaptic neuron fires, creating a new spike that has the potential to travel further along the neuronal chain.[73]

The inner self is an electrochemical entity. Millions of synapses percolate continually in each human brain, almost entirely without our conscious participation, producing thoughts and feelings as well as our sense of self. Does that make sense? Good—you've just experienced an electrochemical event. Disagree? Good—you've just experienced an electrochemical event.

Everything about us is stored within neural connection patterns, which, following the terminology of neuroscientist Joseph LeDoux, we'll refer to as circuits. A system is a more complex circuit that performs a specific function, such as seeing, hearing, or reacting to danger. Circuits transmit their signals to convergence zones, whose job is to synthesize incoming data. The hierarchical arrangement of information processing circuitry, exemplified by cascading convergence zones, is what allows our perceptions to become conceptions, and it's also what enables our internal representations of an external world.

### But Wait—There's More

Patterns of connectivity don't tell the whole story. Our brains are also bathed in chemical "modulators," which include peptides, amines, hormones, and any drugs you might at this very moment happen to be on. Modulators operate diffusely across the brain. Hormones and drugs, for example, which reach the brain via the bloodstream, permeate multiple regions at once.

Suppose you're anxious. You're too busy to meditate or go jogging, so you reach for the Valium. It helps you relax, but apprehension sets back in when you remember that Valium is addictive. You need a drink. Drugs like LSD affect limited brain sites, but alcohol seeps into just about every synapse you've got; your motor control, speech, judgment, and libido go downhill.

You might be thinking wait a minute, my libido actually gets revved up after a few drinks. The good news is that loss of inhibition might spur on your boldness and desire, but the bad news is that with alcohol, "revved up" doesn't always translate to "job well done," an important consideration if you can believe my spam folder. In the original film version of *Arthur*, about a well-heeled inebriate, Arthur (Dudley Moore) is chatting up an attractive woman until she interrupts him to point out that she's a prostitute he had just hired. Arthur replies sorrowfully, "And I thought that I was doing so great with you." Alcohol, like Valium, is a depressant, but with alcohol comes questionable judgment, plus alcohol too is addictive.

By now you're depressed so let me console you. First of all, old Dr. Feelgood is here to help, and second, join the club. One out of ten Americans is currently on antidepressant medication[74], and the World Health Organization predicts that depression will soon rival heart disease as the disorder with the highest global disease burden.[75] After news like that you could no doubt use some Prozac. Prozac was the first of a category of antidepressants called SSRIs (selective serotonin reuptake inhibitors), a class that also includes Paxil, Zoloft, Lexapro and others. Serotonin is a neurotransmitter that's thought to affect feelings of well-being, and SSRIs prevent happy little serotonin molecules, once released into the synapses, from being reclaimed and broken down.

Do SSRIs work? Yes. Do they work better than exercise? Possibly not. Recent studies suggest that regular exercise may be just as effective at mitigating depression, but I'm not sure if the researchers recognized that pill-popping itself provides a great upper-torso workout. Ask a body builder.

A psychoactive drug, which is a medication that affects perception, mood, thoughts, feelings, cognition, consciousness, or behavior, operates at the synapses of your brain. Don't take drugs? High on life? Your inner reality is still governed by identical synaptic mechanisms. Instead of being ruled by inborn chemical dictators, wouldn't you rather elect your own?

## ARE YOU GOOD AT MULTITASKING?

Galvani rightly surmised the electrical nature of nerve impulses, but it would be inaccurate to think of a nerve as some sort of copper wire. Electricity in a wire can travel near the speed of light, but action potentials propagate far more slowly, like a wave of electro-molecular dominoes, attaining speeds no greater than 230 miles per hour (which is about one-third the speed of *sound*). Plus, each nerve pulse can travel only a limited distance, and transmission errors may occur via defect or chance.

Given how creakily its individual circuits operate, how can the brain beget the rich, complex, and clever personalities embodied in the likes of you and especially moi? It's because at any moment thousands of circuits and systems, each fulfilling a specific function, are firing away. The brain resembles a parallel processing computer, which by definition would make it good at multitasking, but are "you" good at multitasking? Clearly, that depends on who and what you think you are.

If you're prepared to identify yourself as the millions of neural micro-calculations that occur every second outside consciousness, then you're good at multitasking. But if you experience yourself as the inner air traffic controller, the overseer who sits in the conning tower, at point zero, managing perceptions, thoughts, and feelings as they incessantly circle, then you're bad at multitasking.

The reason you're able to drive a car and carry on a conversation at the same time is that driving has become ingrained in your neural circuitry. It's automatic, which means you've learned the task so well that you don't have to focus the spotlight of your attention on it. You'll snap-to quickly, and clam up, if another car comes careening through the intersection.

Attention is the key. You can consciously focus on only one thing at a time. If you're performing concurrent tasks, your brain must move the focal spotlight from task to task, dropping the thread of the first and reorienting back into the second, all of which burns up neural resources. "Tech jugglers" are people who simultaneously surf the web, answer emails, watch TV, tweet, talk on the phone, and all the other stuff you may be doing right now. Studies show they accomplish less than people who do one thing at a time. So sure kid, you can put music on while you study, as long as you never actually listen to the music. And FYI, kids will say that music helps them concentrate, but it doesn't; what it really does is provide intermittent relief from the drudgery of studying.

How would you prefer to identify yourself after all? Is your true self a labyrinth of probabilistically firing synapses that operate below awareness, or would you rather stake your identity on the directorial voice in your head? Perhaps both? No matter—you've just experienced an electrochemical event.

## NATURE, NURTURE, LEARNING, AND MEMORY

Question: What do nature, nurture, learning, and memory have in common? Answer: As far as the brain is concerned, they're the same. Each works by tweaking neural connections; each relies on the brain's plasticity.

First, let's put the nature versus nurture debate to bed. The jury is in. Not only does a Matterhorn of evidence suggest that both sides are correct, it shows that overall the split is 50-50. The game ends in a tie. No sudden death, no penalty kicks, not even a seven-point tie-breaker. It's Miller time. Genes construct and preconfigure our neural networks, which is why babies are born with a unique personality, and then the connections are modified by experience.

The relative contribution of nature versus nurture varies among specific mental and behavioral traits, but across all characteristics it's close to 50-50. Here's an example from neuroscientist Joseph LeDoux regarding how nature and nurture contribute to schizophrenia.

> Schizophrenia occurs in about one percent of the general population. In contrast, in identical twins (100 percent gene overlap), if one child has schizophrenia then there's roughly a 50 percent chance the other will develop it at some point. But in fraternal twins (50 percent gene overlap), the likelihood drops to 17 percent. In siblings (25 percent gene overlap), the figure falls to 9 percent, and in first cousins (12.5 percent gene overlap) to 2 percent. The overall picture that emerges is that schizophrenia is strongly tied to genetic factors. That is, in genetic terms, there's 50 percent concordance among identical twins. But one can view this cup as half full or half empty— there's also 50 percent discordance. If genes fully "explained" schizophrenia, concordance would be 100 percent.[76]

Plasticity, as mentioned, is the general term for our brain's ability to reconfigure its neural connections, which allows us to adjust to our environment over time. Learning and memory represent plasticity in action, and they're the same.

When Hebb proposed plasticity in 1949 he had no way to detect the underlying neurochemical mechanisms, but we realize now that learning, and the memory that retains it, are the result of molecular changes in our synapses. When adjacent neurons fire, electrochemical adjustments occur on either side of the synaptic cleft. Typically, changes are undone and the synapses return to their original state, but if a circuit fires often enough, or with particular extenuating support from other circuits, there can be long-term signal-enhancing modifications to the molecular structure of each synaptic membrane. Firing together wires the neurons together, and we find we've learned something or formed a memory.

An explicit memory is something that we can verbally recall, like "I just drank a bottle of Charles Shaw's finest, which I bought at Trader Joe's." An implicit memory is one that lingers below consciousness, such as how to ride a bicycle. Once something is implicitly learned we don't forget it, and we needn't think about what we're doing. When first learning to ride a bicycle, of course, it's necessary to step through the protocol with explicit care. (Mount the bike, push off with one foot, start pedaling, stabilize the handlebars, and so on.) The same is true for learning to drive a stick shift. (Foot off the gas, depress the clutch, shift, release the clutch, etc.) And it's true for learning sex as well, but I'll leave the play-by-play, as well as any color commentary, to you.

## WINDOWS ON THE BRAIN

For Alzheimer's victims, one of the first things to go is short-term memory, because Alzheimer's tends to strike first in the hippocampus, a neural convergence site for working memory. The result is anterograde amnesia—sufferers can't form new memories.

### Amnesia in Film

Anterograde amnesia is featured in the Christopher Nolan movie *Memento*, where the main character, played by Guy Pearce, faces two problems: 1) he has to unravel a murder in which he was involved but of which he has no other recollection, and 2) he can't remember anything for more than three minutes. As you can imagine, or as you know if you've seen the movie, he takes a lot of notes.

Less frequent in real life but more common in film is retrograde amnesia. That's where a person can't remember anything that happened prior to a cranial insult, which in cinema is usually some type of blow to the head. Retrograde amnesia is rare because long-term memory is distributed across more regions of the brain, which also explains why childhood memories are often the last to go in Alzheimer's. Though infrequent, retrograde amnesia is so handy in constructing a mystery that it's been featured in a number of films, but the one that just popped into my mind is David Lynch's erotic whodunit *Mulholland Drive.*

Normally, I look up or even re-watch a movie before I mention it, but this time, because we're discussing memory, I decided not to do that. Rather, I'll describe my honest recollection of the film, which I last viewed several years ago. I'll fill in the missing information shortly.

Here then is a plot summary of *Mulholland Drive* as I recall it. A voluptuous woman (actress to be filled in momentarily) wanders away in tatters from a car crash, eventually stumbling into a house on Mulholland Drive. Inside the house she encounters several people who discover that the woman can't remember who she is or what's happened to her. One of the others is another beautiful woman (Naomi Watts) who wants to help her. The two women have a sex scene together. The End.

I know, I know, there must be other plot points, and surely the mystery of who the woman is and what happened gets resolved, but until I look it up I won't be able to tell you. What I *can* tell you is that I've inadvertently illustrated three of the cardinal rules of memory, which I'm not making up: 1) you remember something better if you already have a memory with which to associate it, 2) you remember what you pay attention to, and 3) memories are more strongly encoded when accompanied by an emotional response. With respect to point #1 (prior knowledge), I didn't recognize the first actress but I already knew Naomi Watts and David Lynch, so those names stuck with me. Regarding points #2 and #3 (attention and emotion), what can I say?

Here's the missing information. *Mulholland Drive* was released in 2001 and the amnesiac was played by Laura Harring. If you're interested in the forgotten plot points, below is a complete synopsis taken from IMDB.com, which now that I read it might help you forgive me for losing the thread of the storyline.

> A bright-eyed young actress travels to Hollywood, only to be ensnared in a dark conspiracy involving a woman who was nearly murdered, and now has amnesia because of a car crash. Eventually, both women are pulled into a psychotic illusion involving a dangerous blue box, a director named Adam Kesher, and the mysterious night club Silencio.

You might be asking, "what's your point?" My point is that while some neural circuits are distributed across the brain, and despite the fact that plasticity allows for a degree of rewiring, specific mental functions tend to be localized within specific regions of the brain. That's why, for example, damage to the hippocampus causes the short-term memory loss associated with anterograde amnesia. Mapping brain function to brain location is a major effort of neuroscience.

Originally, the only option was to note what happened after some poor sap got whacked on the head. One famous study involved railroad worker Phineas Gage, who in 1848 accidently set off an explosion while tamping blasting powder into bedrock. The three-foot long, inch-round tamping iron rocketed through his jaw, out the top of his head, and landed 100 feet away. Miraculously, Gage survived. He began walking and talking within minutes, was first treated then studied by doctors, and lived another twelve relatively normal years. His personality had changed, though, and the correlation between those changes and his injury afforded neuroscientists one of their first opportunities to map brain function to its location. Today, Gage's skull and the tamping rod that shattered it remain on permanent display at Harvard's Library of Medicine.

In the mid-twentieth century, advances in brain surgery allowed doctors and animal researchers to more directly assess the relationship between brain damage and cognitive functioning. Starting in the 1980s, less intrusive technologies such as brain imaging emerged. Magnetic resonance imaging (MRI) and computed tomography (CT) are suited for structural or static imaging, where brain changes of interest, including tumors, occur over months or years. MRIs provide better pictures of soft tissue, but CT scans are faster and cheaper, so they're commonly used in hospital emergency rooms. The encephalographic techniques, electro-encephalography (EEG) and magnetoencephalography (MEG) operate on millisecond time scales, providing graphs of rapid-fire numeric brain state data, but no images. EEG is where they paste the little electrodes to your scalp, and MEG is where they call in Anton Mesmer. Positron emission tomography (PET) and functional magnetic resonance imaging (fMRI) can now resolve successive images within seconds or less.

---

*Have You Ever Had an MRI?*

Magnetic resonance imaging is one of our most informative diagnostic tools, but you deserve to know that MRI machines are manufactured with parts recycled from World War II era U-boats. First they trolley you into a repurposed torpedo tube, then they fire up the engine room. Clang, bang, thank you Wolfgang.

fMRI, which produces a stop-motion animation of brain activity, has been the tool of choice for many researchers since the early 1990s. fMRI has been used clinically to investigate hemispheric asymmetry in language and memory, map the neural correlates of a seizure, study how the brain responds to a stroke, examine the effect of drugs, and detect the onset of Alzheimer's and depression.[77]

Could I read your mind? I don't mean could I, in theory, read your mind, I mean would you be willing to get into the fMRI machine and answer a few questions? One fMRI study examined people's neural response to real Coca Cola versus the same Coke in an unlabeled bottle (they reacted differently). Another study examined the brain activity that characterizes men's preference for sports cars, and there was one that looked at differences between Democrats and Republicans in their reaction to images of the 9/11 attacks.[78] fMRI studies have shown that men and women respond differently to chocolate, and that chocolate can activate many of the same brain regions as heroin, cocaine, or sex. Of course, chocolate is better than sex because it won't upset coworkers if you have it on your desk.

Does that strike you as funny? You don't have to answer. If you'll just slide into the fMRI machine I can take a peek for myself, and as a matter of fact, I'll know your response before you're aware of it. I'll be able to see what you can't: basal circuits performing low-level synaptic calculations and passing their tallies up a chain of convergence circuitry, until the grand totals—your answer—pops into consciousness.

Recently, Taiwanese scientists utilized fMRI to inspect the brain patterns of humor, and concluded that, "our findings provide a better understanding of the neurological mechanisms of incongruity detection and resolution during humor comprehension. Extended studies in this topic could be conducted by comparing the sex differences in the neural correlates of humor comprehension."[79] Gender differences in humor comprehension? Never noticed.

### FROM A COMPLEX BRAIN TO MY SIMPLE MIND

Do you have good days and bad days? Sunny days then cloudy days? A day of rainbows chased off by a typhoon named Beverly? Then your brain, like the weather, just might be a "complex adaptive system," and in fact that's what some scientists would tell you. They contrast adaptive systems with determined systems, where cause and effect are foreseeably and reproducibly connected. Determined systems are often linear as well, meaning that effects vary in proportion to their cause. For instance, we all hope that our car's steering system is fully determined—predictable, reliable, and proportionally responsive to our movements. Weather, of course, is notoriously unpredictable, even when powerful computer models crunch every bit of obtainable data, which is why the only accurate long-term forecast turns out to be "intermittent rain."

A complex adaptive system is a network of semi-autonomous agents that follow simple local rules (not necessarily perfectly) as they interact with each other and their environment. Complex systems may cycle through a familiar set of configurations, but the order and scale of individual states can vary unexpectedly, and the systems tend to exhibit nonlinearity, where marginal differences in initial conditions elicit wide variation in the final figurations. Weather's local agents are atmospheric molecules, which when energized by solar radiation and Earth's rotation interact to produce the patterns we call weather.

In the brain, billions of semi-autonomous neurons continually percolate, firing or not along one or more neural pathways. The brain's massively parallel electrochemical architecture, already indeterminately complex, is also susceptible to incidental nano-events and microscopic errors. The system's final outputs, such as the likes of you and me, can be unpredictable to the point of charm.

When we watch a flock of birds in flight we're watching a complex adaptive system in action. Starlings, for example, despite the appearance of choreographed grace, boast no prima ballerina, no leader. Each bird, and there can be thousands, follows three local rules: avoid hitting neighbors, align flight with neighbors, and fly a prescribed distance from neighbors. From these simple rules, instinctively but imperfectly applied by each bird, emerges the fickle swooping and turning of the flock.

Scientists and philosophers who regard the brain as a complex adaptive system contend that both consciousness and self are "emergent properties" of the brain, meaning that consciousness and self cannot be reduced to neuronal agency. Just as we can't foretell weather based on the interaction of air molecules, we'll never understand consciousness and self by studying the activity of neurons.

The implications of a complex systems brain model are notable, especially with respect to free will, intentionality, personal identity, and the autonomy of self, thorny issues we'll take up later, but for now it's enough to recognize that complex adaptive systems offer an intriguing perspective from which to ruminate over some of life's big questions.

The subject of this book is moi, a rare bird indeed, and trying to decrypt its singular soaring tango can be tricky. Or perhaps each of us is just a Rockette in some cosmic chorus line, briefly on loan from central casting. Science and informed reasoning, let's hope, can help us sort through the matter and demystify self. According to Einstein, "If you can't explain it simply, you don't understand it well enough." That's easy for *him* to say.

# THE MIND

*"What we call a mind is nothing but a heap or collection of different percep-*
*tions, united together by certain relations and supposed, though falsely, to*
*be endowed with a perfect simplicity and identity."*
*- David Hume, philosopher*

*"We need to see our defective Stone Age minds for what they are if we*
*ever hope to drag ourselves, kicking and screaming, into the twenty-first*
*century."*
*- Hank Davis, evolutionary psychologist*

*"The notion that we have limited access to the workings of our minds is*
*difficult to accept because, naturally, it is alien to our experience, but it is*
*true: you know far less about yourself than you feel you do."*
*- Daniel Kahneman, Nobel psychologist*

---

## *FROM BRAIN TO MIND*

Are you your mind, or do you have a mind? If you are your mind, how could you know if you had one? And if you have a mind, what's your relationship with it? Get out your beekeeper's suit because we've entered a hornet's nest of intellectual discourse known as philosophy of mind, where, as George Berkeley noted, "philosophers kick up the dust and then complain they cannot see." Ironically, attempts to decipher philosophy of mind can be hazardous to one's sanity, so we'll stick with science.

When medical doctor Daniel Siegel surveyed 100,000 mental health professionals about whether they had ever witnessed a lecture defining mind, 95% of them reported that they'd never attended even a single lecture where mind was defined.[80] We'll adopt a broad definition in which mind encompasses consciousness, thought, perception, will, memory, emotion, reasoning, mental imagery, attitudes, desires, beliefs, and imagination.

It's tough to sell ourselves on the notion that we don't know our minds. There's plenty of scientific evidence, but for each of us there's only one unequivocal expert, the genuine insider—me. I'm in here, I can see what's going on, and I know why. The reality that we operate largely on automatic pilot seems both unintuitive and unappealing, another reason we'll stick with science, and what scientific research shows is that we're masters of self-deception. This is how psychologists Christopher

Chabris and Daniel Simons view our situation: "We all believe that we are capable of seeing what's in front of us, of accurately remembering important events from our past, of understanding the limits of our knowledge, of properly determining cause and effect. But these intuitive beliefs are often mistaken ones that mask critically important limitations on our cognitive abilities."[81]

Much about mind is up for debate, but I see no way to get around one initial fact: brain generates mind. Psychologist Steven Pinker is among many who support the adage that "mind is what the brain does," a contention affirmed by brain injury cases, animal experimentation, brain surgery, and drug incidents, all of which have demonstrated how aberrations in the brain can affect cognition, emotion, and motivation, the three classic facets of mental life. If you're unconvinced I'll bet that I can change your mind, literally; just loan me your skull along with a cordless drill and a swizzle stick.

The hottest topic among philosophers and cognitive scientists is whether there's any attribute of mind that isn't a direct neurochemical artifact of the brain. Does mind simply reflect neuronal patterns or is it something more? There are those who posit that mind is an emergent property of a complex system brain, inherently and forever inexplicable by neural mechanisms, to which their critics respond that emergence "raises the specter of illegitimately getting something from nothing."[82]

---

### What Do Margaritaville and the Vatican Have in Common?

The aforementioned Dr. Daniel Siegel believes that psychological integrity is more than the sum of its parts. "Integration is more like a fruit salad with heterogeneous elements rather than a smoothie that has been blended into a homogeneous mixture."[83] As I understand him, a fruit salad retains its individual components yet is also overlaid by an emergent salad-ness, whereas a smoothie has no parts to be greater than. Now, I whip up a smoothie just about every morning, and unless Dr. Siegel's blender has a setting beyond "liquefy"—maybe "obliterate" or "kablooey"—parts remain. Siegel founded an interdisciplinary approach called interpersonal neurobiology, a model with scientific and humanistic merit, but it's definitely weak in the area of food preparation.

Let's test our understanding of Siegel's perspective: Is a margarita something greater than the sum of its ingredients? Correct answer: On the rocks, yes (parts exist), frozen, no (parts pulverized). Most common answer: Yes, right now a margarita would be great.

Is there more to the mind than refracted brain states? Most scientists say no. Some, like Siegel, say yes. Others say "Amen." Pope John Paul II must have seen science's handwriting on his medieval city walls, because in 1998 the Vatican invited clergymen and scientists to a conference on "Neuroscience and Divine Action," the goal of which was to reconcile how God could touch people without manipulating their neurons or breaching the laws of physics. Neuroscientist Joseph LeDoux, who was in attendance, reported back that "the Vatican conference ended inconclusively."[84] Attendees might have reached consensus given worthier experimental equipment, like a blender full of margaritas.

Our bodies interact with the world around us, but our minds never do. What we regard as the outside world is a kind of virtual reality constructed by the brain and implanted into our minds. We experience the physical world via our senses, a process we usually call perception, but eyes can't see and fingers cannot feel. What we discern is what the brain synthesizes from nerve impulses that have been passed to it. The world of subjective experience is brain-created.

### The Truth Is Out There

"OK, but at least it's my world; le monde de moi."

By what reckoning? Did you create your brain? Do you operate it? Are you operating it right now?

"It must be my brain, my mind, because it's in my body."

Are you sure? Just because we happen to find ourselves in the center of something does that make it ours? Do people figure, "this must be my mountain meadow, because here I am standing in the middle of it"? That's ownership by squatting, which real estate law refers to as adverse possession. In any case, whether or not we're corporeal squatters, we'll soon enough quitclaim all our property, real or imagined, including our selves.

And let's not confuse adverse possession with demonic possession. In adverse possession some entity—let's call it [fill in your own name here]—tries to assume ownership of the premises based on discovering itself there. Demonic possession on the other hand is characterized by ungovernable voices in the head. If you experience uninvited cranial chatter, unsolicited thoughts or feelings, here's a test to check if it's a demon. Try spinning your head around 360 degrees or throwing up on a priest from across the room. If you can't perform those feats, you have a more common condition that doctors describe as "running off at the brain."

There are people with fully functional eyes and fingers who are unable to see or feel due to cerebral defects. According to the Mayo Clinic, two of the many complications that can arise from brain trauma are loss of vision and loss of facial sensation[85], and there's a hereditary condition called congenital analgesia in which a person can't feel physical pain. The disorder makes for a difficult and bizarre life, dangerous for its victims but so tailor-made for TV melodrama that it's been featured on episodes of *Grey's Anatomy*, *House M.D.*, and *The X-Files*.

And here I could use some help. If you've seen *The X-Files* you might recollect that the dialogue between Fox Mulder and Dana Scully (David Duchovny and Gillian Anderson) tends to be terse, direct, and stylized. They call each other Mulder and Scully, and their repartee bounces back and forth like a ping pong ball. For fun, I made up some *X-Files*-style dialogue that features the inability to feel pain, hoping that there might be a respectable joke there. I came up with a setup line and several punchlines, but I'm not sure they work. I invite you to choose the punchline you like or invent a better one.

### (*Made-up Dialogue*)

Mulder:  Scully, you're a doctor. Why can't this man feel anything?

Scully:  Mulder, he's a politician.

Scully:  Mulder, he just watched *Keeping Up With the Kardashians*.

Scully:  Mulder, this is a morgue.

More important than which punchline you picked is that you read through them. Or did you? Here's how your vision is working at this precise instant. Your eyes are moving across the page, taking in words, but what you don't notice is that you're not actually reading all the words (which is why we never catch every typo), that your peripheral vision is smeared and colorless, that you have two blind spots the size of tennis balls at arm's length and off-center from your face, and that every time you move your eyes your vision is momentarily blacked out (to eliminate blurriness).

Our innate perceptual camerawork is as shaky as *The Blair Witch Project*, but the brain, with its Hollywood editing skills, assembles the raw footage into a smooth panoramic pageant for us to behold. The other sensations that brain presents to mind are likewise fabricated. As neuroscientist David J. Linden puts it, "In the sensory world, our brains are messing with the data."[86] Experimental psychologist Bruce Hood goes even further: "The same deception is true for all human experience, from the immediacy of our perception to the contemplation of inner thoughts, and that includes the self."[87]

Have you ever glanced at the second hand of an analog clock, or the readout on a digital clock, and you could swear it had momentarily quit moving? It's called the stopped-clock illusion (chronostasis), and if you'd like to check it out for yourself here's how and why it happens. Chronostasis can occur after you've swept your eyes across a visual field and brought them to rest on a clock, preferably one that runs silently. After eye movement the brain must reconstruct your visual reality, and during that lapse the clock doesn't move, at least for you. Similar lags have been demonstrated in tactile and auditory perception, though to test out auditory delay you'd need a phone with a separate handset, so we can probably forget about that.

The brain constructs our inner world, which includes a model of external reality and one of the self. Aspects of the simulation are passed on to the conscious mind, but it's important to remember that natural selection has never cared whether our minds experience truth or illusion. What matters are mental characteristics that foster reproductive fitness. Most scientists agree, because there's plenty of evidence, that we've been designed to look at the world through self-serving rose-colored glasses, but I suggest it might be more helpful, reproductively speaking, to doff the glasses, head down to the local bar, and don some beer goggles.

## FLINTSTONES FAMILY VALUES

The central question of philosopher David Hume's 1740 book, *A Treatise of Human Nature*, asks, "Why is human nature what it is?" Hume was an empiricist, but rather than providing tidy answers his book suggests how we might scientifically approach the question, which is exactly what neuroscience and psychology are doing today—examining human nature through the lens of evidence. Four generic agents have been proposed to account for human behavior: nature (genes), nurture (upbringing), culture, and self-control, and while it's indisputable that each individual's personal disposition is affected by upbringing, culture, and life choices, Hume was interested in the ubiquity of human nature. For instance, all cultures exhibit love, ambition, language, sexual desire, marriage, art, grammar, music, and smiles, and in all societies men kill more frequently than women, twenty-year-olds are considered more physically beautiful than the elderly, wealth can purchase power over others, people discriminate in favor of their clan and against strangers, and parents love their children.

That's where the field of evolutionary psychology (EP) comes in. When evolutionary psychologists find a cross-cultural trait they attribute the characteristic to human adaptation via natural selection, which is to say that the trait must provide a reproductive advantage to those who possess it.

But I misspeak, because while human culture arose within the last 100,000 years, and civilization within the last 12,000, our brains and the minds they engender evolved over millions of years. What I should have said is that a mental trait must *have provided* a reproductive advantage to those who *possessed* it. As human encephalization led to accelerated cognitive adaptation, it did so within a prehistoric environment, which is the reason evolutionary psychologists like Hank Davis wave a red flag whenever our Stone Age minds confront the modern world, which of course is every day. We know that experience can rewire circuitry in our cortex, but we're also aware that our midbrains and hindbrains are not affected by life events (with the exception of physical trauma). Do you honestly believe that the jealousy you feel when your lover flirts with somebody else is a rational, justified, productive, and healthy response to the situation? Get over yourself—it's Pleistocene reckoning.

Evolutionary psychology is controversial. EP proposes that our thoughts, feelings, and behavior stem from antiquated contingencies, but the specific hypotheses that EP puts forward are largely untestable. How can I prove that you're jealous because of Stone Age brain processes and not because of that time your ex ran off with a troupe of Swedish gymnasts? I can't prove it, and that's why EP sometimes gets a bad rap. Neural circuits don't leave fossil imprints, so we can't compare them to contemporary brain images.

David J. Buller is a philosopher of science who regularly impugns evolutionary psychology. One of his beefs with EP, which he lays out in his book *Adapting Minds* and again in a recent *Scientific American* article entitled "Four Fallacies of Pop Evolutionary Psychology," is EP's standard proposition that men become sexually jealous while women become emotionally jealous. According to the prevailing EP doctrine, men can't tolerate the thought of a woman sleeping with another man, whereas women can't bear the idea of a man caring for another woman. Buller's own theory of jealousy, called "the relationship jeopardy hypothesis," maintains that men and women each possess the same evolved capacity to distinguish when a relationship is threatened.

Science fiction writer Theodore Sturgeon, in defending the overall mediocre quality of science fiction, coined an adage that now bears his name, Sturgeon's Law: Ninety percent of everything is crap. Perhaps the law applies to evolutionary psychology, a field in which it's difficult to implement standard scientific methods, but to critics of evolutionary psychology (and also critics of analytic philosophy, sociology, cultural anthropology, macroeconomics, plastic surgery, improvisational theater, television sitcoms, philosophical theology, massage therapy, and other disciplines), philosopher and cognitive scientist Daniel Dennett suggests "Don't waste your time and ours hooting at the crap! Go after the good stuff or leave it alone."[88] Richard Dawkins too states that evolutionary psychology is "unjustly maligned."[89]

According to theoretical physicist Leonard Mlodinow, coauthor of *The Grand Design* with Stephen Hawking, "We think of ourselves as a civilized species, but our brains are designed to meet the challenges of an earlier era."[90] And lastly, it's worth noting that dozens of universities, including Harvard, Stanford, and Cambridge, currently offer coursework in evolutionary psychology.

My personal experience is that whenever two reputable factions dispute an issue in good faith, the truth lies somewhere in the middle. Science invites argument, and it's in the crucible of dissent that truth sometimes crystallizes. To me, it seems logical that if the brain evolved via natural selection, and if mind emerges from the brain, then mind must be to a great extent the product of natural selection. Evolutionary psychology probes into the details, where of course dwells the devil.

For Charles Darwin, natural selection is blind to good and evil— it's a numbers game pure and simple—if I produce more copies I make more people like me. We're born with physical and psychological traits than can then be modified by experience; evidence suggests that physical characteristics may be easier to refashion than personality.

---

### American Muscle

Consider the actor Christian Bale, who normally weighs about 185 pounds. For *American Psycho* (2000) he worked out three hours a day for several months. He retained his pumped-up physique for the dragon flick *Reign of Fire* (2002), where he fought shoulder-to-shoulder and abs-to-abs with Matthew McConaughey. For 2004's *The Machinist*, Bale dropped to 120 pounds, at which point the producers commanded him to stop losing weight. If you've seen the movie you know he looked like a skeleton. A year later Bale made *Batman Begins*, wherein he flaunted 190 pounds of chiseled muscle. His weight remained more or less normal until *American Hustle* (2013), where he put on fifty pounds of certified beer belly. David O. Russell, the movie's director, told the scientific journal *Us Weekly* that Jennifer Lawrence commented, after kissing Bale, "I finally get to make out with Christian Bale and he's a really fat guy. He's Fatman, not Batman."

Physical characteristics are not set in stone, and likewise, mental proclivities are not set in the Stone Age, unless we're unaware of them, deny them, or are unprepared to confront them. So let's face up to some EP propositions about mating. As mentioned, males tend to be insecure about sexual infidelity and females about emotional infidelity. Since males can never be certain about paternity (unless they demand DNA testing, which rarely happened in the Pleistocene era), it's genetically advantageous to guard a mate from other males. Male jealousy evolved as the emotion to enforce that strategy. And since females require sustenance and protection to raise children, it's to their genetic advantage to have a resourceful mate who sticks around. Voila, female jealousy.

One implication of the bifold priorities, according to EP, is that females will be choosier than males in mate selection, and another is that they'll favor prosperous men (as Henry Kissinger put it, "power is the greatest aphrodisiac"[91]). Males, says EP, will favor the most fertile females, the young and nubile. If any of that rings a bell, welcome to the provocative world of EP, and welcome to your authentic former cave life. According to evolutionary psychologists we're guided by the genes of grotto dwellers and not by rationality, and on more occasions than we realize, Stone Age fixations win out over New Age intentions.

## I'M OF TWO MINDS ABOUT IT

The devil made me do it. The spirit is willing but the flesh is weak. Said Plato, "First the charioteer of the human soul drives a pair, and secondly one of the horses is noble and of noble breed, but the other quite the opposite in breed and character. Therefore in our case the driving is necessarily difficult and troublesome."[92] I don't know what came over me. I'm not myself today. According to Geraldine (comedian Flip Wilson in drag), "The devil made me buy this dress."

We all experience bouts of inner conflict, yet what we're talking about here is not the long-debated tug-of-war between body and soul or between reason and passion, but the foundations of such incongruence, which reside in nonconscious mechanisms of mind. (I should mention that some contemporary scientists prefer to speak about nonconscious processes instead of unconscious ones in order to distance themselves from Freud.) We're talking here about human beings' well-documented dual-process cognitive system. Research dating back to 1970 suggests that our brains support two different modes of cognition, one conscious and the other nonconscious. We are literally of two minds. Cognitive psychologist Arthur S. Reber, best known for developing the concept of implicit learning, stipulates that consciousness was a late evolutionary development, and that the nonconscious mode is our predominant and default mental system. According to Reber (and others), the notion that consciousness is our primary system is an illusion fostered by ourselves and other people.

---

### Tennis Anyone?

Suppose you've been invited to compete in a Legends of the Game tennis tournament. You're crouched in the deuce court, a couple of feet behind the baseline, awaiting serve from Pete Sampras. From 26 yards away, Sampras blasts a 130 mile per hour bullet down the T. Within half a second you sense the ball, pivot to the left, pull back your racquet, and drive a topspin backhand down the line, right past the charging Sampras. Even Sampras applauds quietly in appreciation. Hey, it could happen. Especially if you happen to be Andre Agassi, one of the great returners of all time. My point is that several top pros say they're tracking the ball before they see it.

---

Seeing without awareness is called blindsight, and it utilizes our secondary, primitive, nonconscious visual system. Subjects in blindsight research are almost always cortically blind; their eyes are fine but their visual cortex is damaged, at least on one side of the brain, and they can't see. In a typical experiment, stimuli are presented to the blind eye and subjects are asked to guess what the object is. Even though they swear they can't see the object, their guesses always exceed chance, often with 70% accuracy.

A recent experiment involved a subject known as TN, who in 2003 suffered grave misfortune when the primary visual cortices on both sides of his brain were destroyed by strokes. TN could see nothing, not even large objects coming straight toward him, yet in 2008 researchers who suspected TN might exhibit blindsight convinced him to walk down a hallway without his cane (an attentive researcher would walk directly behind him). They told TN the hallway was empty, but in truth it was littered with objects. TN successfully navigated around a wastebasket, a paper shredder, a tripod, some small shelves, and a cardboard box. He said he had no idea how he did it. It's a famous experiment that you can watch by searching for "blindsight" on Youtube. The man you'll see walking right behind TN is Lawrence Weiskrantz of Oxford University, who during the 1970s, in the face of widespread skepticism, proposed the existence of blindsight.

Just FYI, you can view two other famous psychology experiments on Youtube by searching for "Daniel Simons." The first is the door study that I described in the book's introduction, where a direction-seeking experimenter switches places with a second experimenter who's carrying a door. If you don't believe a subject could fail to notice that he's talking to a different person, you can see it for yourself. The second experiment, by Daniel Simons and Christopher Chabris, involves selective attention. Subjects are shown a video of players passing a basketball around and asked to count the passes made by white-clad players, and while many subjects count accurately, almost none notice a gorilla walking among the players. Many participants refused to believe they'd missed a gorilla, even after Simons showed them the video again so they could see it for themselves. They protested it was a different video.

Blindsight is the primitive hidden half of a dual-process system for vision. Most neuroscientists and cognitive psychologists now believe that human beings possess a dual-process system for almost all cognitive activities, with each system serving a different function and having its own strengths and weaknesses.

Dozens of scientists have laid out dual-process models, including Nobel laureate Daniel Kahneman in his award-winning book *Thinking, Fast and Slow*, but in all cases each half of the system includes similar characteristics. We'll borrow the somewhat nondescript terminology of cognitive scientist Keith Stanovich and refer to the systems as System 1 (automatic processes) and System 2 (analytical processes). Below are the characteristics of each modality, based on a list developed by researchers Keith Frankish and Jonathan St. B. T. Evans.

|              SYSTEM 1               |              SYSTEM 2             |
| ----------------------------------- | -------------------------------- |
| Unconscious, preconscious           | Conscious                        |
| Evolutionarily old                  | Evolutionarily recent            |
| Implicit knowledge                  | Explicit knowledge               |
| Automatic                           | Controlled & effortful           |
| Shared with animals                 | Uniquely human                   |
| Fast                                | Slow                             |
| Parallel                            | Sequential                       |
| High capacity                       | Low capacity                     |
| Intuitive                           | Reflective                       |
| Contextualized                      | Abstract                         |
| Pragmatic                           | Logical                          |
| Associative                         | Rule-based                       |
| Independent of general intelligence | Linked to general intelligence   |

In *Thinking, Fast and Slow* Kahneman refers to System 1, because it does most of the work, as "the hero of the book," then adds that we'd prefer to believe, and do believe, that System 2 is our inner hero. System 1, says Kahneman, automatically and diligently perceives and processes inputs, sorts through feelings and memories, and makes "suggestions" (often via feelings or hunches) to System 2. At that point, or so it seems to us, System 2 produces a decision.

The 1-then-2 approach is fast, computationally efficient, and often serves us well, but System 1, which Kahneman describes as "a machine for jumping to conclusions," exhibits deep-seated biases that System 2 typically can't detect. Our analytical system is slow and effortful, and simply can't pore over each of the thousand minute decisions we make every day, so the dual systems wind up compromising. We often get things a bit wrong, and we occasionally get things precariously wrong.

My favorite dual-system metaphor comes from social psychologist Jonathan Haidt, who tells us that: "The mind is divided, like a rider on an elephant, and the rider's job is to serve the elephant. The rider is our conscious reasoning—the stream of words and images of which we are fully aware. The elephant is the other 99 percent of mental processes— the ones that occur outside of awareness but that actually govern most of our behavior."[93]

The elephant is powerful and stubborn, and though the rider's attempts to control it are not valueless, they often fail. Haidt is among those who maintain that reasoning and conscious planning are recent evolutionary innovations, like new software that he calls Rider version 1.0, whereas our automatic programs have been through thousands of product cycles. Consciousness, says Haidt, evolved later on because it adaptively served the elephant's needs.

In dual-process reality, as soon as a stimulus impinges on our senses the hidden gear-work of System 1 kicks in: should we approach, evade, or ignore; perhaps it's an opportunity, perhaps it's a threat, or perhaps it's nothing. It's not hard to understand the evolutionary merit of a built-in appraisal system like that. System 1 rapidly evaluates the situation based on long-established best-guess heuristics, constructs a bottom-up judgment along with an emotional response, and passes the full mental-emotional package along to consciousness for final analysis. For reasons we'll describe shortly, System 2 usually rubberstamps System 1 then kicks back.

---

### System 1 vs. System 2: The Proof Is in a Thought Experiment

Let's say you're approached at a party by a provocative stranger. Really? When was the last time you were approached at a party by a provocative stranger? System 1 wanted to believe me, or at least indulge me, or more likely indulge itself, so it passed that piquant but dubious tidbit on to System 2, which promptly bought into it. You sidled right up to the notion of yourself and a provocative stranger, all logic to the contrary. Dual process theory says that we will, in fact, dream on.

---

Daniel Dennett and others have described System 1 as a highly parallel distributed processing system and System 2 as a fabricated serial processor, consciousness, a virtual machine that perceives the world as linear. The virtual-system view is a form of dual-process theory in which the second system emerges from the first instead of being distinct from it. Evolutionarily, this viewpoint could explain how a radically new form of cognitive activity, consciousness, could arise without massive changes to our neural hardware.

A final cognitive dichotomy exists between the experiencing self and the narrating self (Kahneman calls the latter the remembering self). The experiencing self resides solely in the present, absorbing moment-to-moment experience, like when we're engrossed in a movie or skiing down a mountain. Most of the time, though, we inhabit the narrating self, which elaborates the past and future (to ourselves and others) via thoughts, words, and ideas, as when we afterwards reflect on the movie or chat over hot toddies in the ski lodge.

## I'M LAZY, BIASED, AND OVERCONFIDENT

If a train leaves St. Louis at 8:00 AM traveling 60 miles per hour ... how far is it to the bar car? I'm lazy because analytical thinking is hard work, whereas when I'm on automatic pilot, System 1's highly efficient parallel processing machinery steers me along with virtually no mental effort. When System 2 gets drummed into action to collect data, analyze a situation, or make a rational decision, I must literally stop and think. I hate when that happens. I've got to redirect attention to the chore at hand then grind methodically forward. Like most people, I'll probably cease cogitating the instant I come on a passable response.

Which of the following two questions is easier: 1) Reduce the fraction 9/117 to its lowest terms, or 2) What makes Shakespeare a great writer? There's no passable solution to the first question; you're either right or wrong (1/13 is the correct answer), and unless you happen to be Rainman you've got to work it through. The second question has plenty of justifiable answers, and one might well have occurred to you. There's Shakespeare's lyrical prose, his characters, his grasp of human nature, his sense of humor, his dramatic structure, and so on. Amazon hawks dozens of books about Shakespeare, such as *The Meaning of Shakespeare* Volumes 1 and 2, but the only book I could turn up on reducing fractions was *Fractions and Decimals for Fourth Graders*. School kids can reduce a fraction, but how many can author Shakespearean criticism? Yet for many of us the second question is more palatable, easier, because we can be done with it as soon as we come up with a reasonable response. In fact, you might have relaxed as soon as you saw the question because you knew you could fab up an acceptable answer on the spot. No need to think it through unless he calls on me.

I'm lazy thanks to the overhead associated with running a virtual serial mechanism (consciousness) on top of a of a parallel processing architecture, so most of the time System 2 unthinkingly rubberstamps System 1's longstanding intuitions. In cases where further analysis is unavoidable, System 2 often works only to the first defensible solution. It's a computationally economic scheme that's frequently effective but sometimes misses the mark, which is why it pays to acknowledge mind's native two-tier strategy and know when to inspect its trustworthiness.

Starting in the early 1970s, Daniel Kahneman and Amos Tversky pioneered an area of psychology called heuristics and biases. I'm biased because when System 1 automatically invokes an inbuilt predilection, System 2, if it even notices, won't work hard enough to overturn the call. Christopher Chabris and Daniel Simons call our distorted perceptions "everyday illusions," because they're ever-present and because, like the visual illusion of Escher's never-ending staircase, they persist even after we realize something is wrong. Our inherent perceptual fallacies remain stubbornly resistant to change even after we become aware of them. In Jonathan Haidt's lingo, the elephant is impervious to the rider's will.

A cognitive bias is a nonconscious, systematic inaccuracy that predisposes our thinking towards a particular perspective, and scientists have documented over a hundred such tendencies. You may be familiar, for example, with the halo effect, our inclination to judge people across-the-board based on one or a few observations. Teachers are notoriously susceptible to the halo effect in evaluating students. We've previously encountered confirmation bias, the innate urge to interpret information in a way that supports our beliefs, and the base rate fallacy, where we pay excess attention to a few extraordinary cases without considering the baseline frequency of such cases overall. Fear of flying and of terrorism are examples of the base rate fallacy; we pay exaggerated attention to deadly incidents and ignore their exceptional infrequency.

I'm sure you and I would never exhibit any of the following inborn penchants, which I'm not making up. We certainly wouldn't fall prey, for instance, to the cheerleader effect, dressing more attractively when we're around other people. Or to the bandwagon effect, tending to fall in with what our peers believe. Or to the money illusion, valuing money without investigating what it might actually do for us. Or to mindless accumulation, amassing resources beyond our needs. Or to the planning fallacy, where we underestimate how long projects will take. Or to rosy retrospection, in which we remember the past as better than it really was. Or to the Lake Wobegon effect, where our personal attributes and those of our friends are all above average.

All right, so maybe you dress up for others, but are you prejudiced toward one ethnic group over another? Most of us exhibit built-in racial preference, as I found out the hard way. Project Implicit is an international nonprofit network of researchers who investigate implicit social cognition—thoughts and feelings outside of conscious awareness and control. To assess your own implicit attitudes go to ProjectImplicit.org, but be forewarned that the results can be disturbing. The majority of people turn out to have negative implicit associations with groups such as African Americans, immigrants, obese people, the handicapped, and the elderly.

---

### Did I Mention I'm Biased?

In the name of science and personal curiosity I invested twenty minutes to take the African American-European American implicit attitude test, where I found to my unforeseen discomfort that I was among the 70% of respondents who had an "automatic preference for European American compared to African American." I knew going in that research consistently confirms the ingroup bias for everything from race to ad hoc work teams, yet I expected to somehow be exempt. I wasn't, and I suppose that's the point. Reflexive impulses gyre along, leading us on, but we don't see them.

---

One of our self-deceptions is called the illusion of skill, and it's displayed most blatantly by so-called pundits, whose job it is to predict outcomes. Two doctoral students at Washington State University, using special software, analyzed the tweets of over a million experts and non-experts as they tweeted predictions about the 2012 baseball playoffs and the 2013 Super Bowl. The students, Wooten and Smith, found that the professionals were correct 47% of the time and amateurs 45% (no better than guesswork in either group). Linguistic analysis revealed that the pros, regardless of track record, were highly confident in their picks.

Daniel Kahneman affirmed the illusion of skill among big-time stock advisors. His account, "Don't Blink! The Hazards of Confidence," was showcased in the *New York Times Magazine* on October 19, 2011. Kahneman had been invited to speak to a group of investment advisors in a firm that provided financial advice to wealthy clients. He requested in advance, and was given, a spreadsheet that summarized investment results for 25 of the firm's wealth advisors over eight consecutive years. Yearly advisor bonuses were based on performance, yet when Kahneman analyzed the data he found no consistency of performance whatsoever. In Kahneman's words, "The results resembled what you would expect from a dice-rolling contest, not a game of skill." Was the firm stunned? Did it quietly close its doors and return investor fees? It was unfazed. En route to the airport, one of the company's executives told Kahneman that he had, "done very well for the firm and no one can take that away from me."

You might wonder why an investment firm allowed Kahneman, a psychologist, to review its books. Did they think he might turn up some new psych-based sales pitch? "Do not ask for whom the stock market bell tolls, it tolls for thee, so start salivating!" Actually they didn't allow him, they eagerly invited him, because, and perhaps I didn't mention this, Kahneman won the Nobel prize in economics (psychology doesn't have one). He and Amos Tversky, who surely would have shared the prize had he been alive, demonstrated that investors aren't rational. They rely instead on sophisticated prediction algorithms such as those built into Chinese fortune sticks and Ouija boards, which is why I don't invest in the stock market (though I'm sure that Mattress of America, where I deposit my money, does).

## Credit Default Swaps Explained

Please allow me to personally demonstrate the illusion of skill by explaining the financial crisis of 2008 in easy to understand terms, by which I mean BS. The financial crisis of 2008 was precipitated by two factors, the first of which was that millions of average citizens defaulted on "securitized mortgages" because they couldn't afford the postage to mail checks to each of the 10,000 investors who owned a share of their mortgage. The second factor was credit default swaps (CDS), a type of "derivative" that has nothing to do with the derivatives found in calculus. Financial derivatives are far more complicated.

Suppose I loan you $1000, which you promise to pay back one year later with $100 interest. I don't quite trust you, for obvious reasons, so I go to the bank and purchase a CDS for $50. If you default, the bank will reimburse me the original $1000; I'll be out $50, and the bank will be out $950. If you pay back the $1100, I'll net $1050 and the bank will have $50. It's like insurance, with a catch—anybody can buy a CDS on your debt. In this case, everyone who knows you buys a CDS, gambling $50 against $950 that you'll default. It turns out that each of the hundred million people worldwide who saw you profiled on *America's Most Wanted* goes to each of ten banks and buys a $50 CDS on your debt. Fifty dollars times a hundred million people times ten banks nets the financial industry fifty billion dollars, which it immediately reinvests. When you of course default, each of the ten banks now owes a thousand dollars to each of a hundred million people—a trillion dollars total— which the institutions don't have because banks are only required to keep $17 in cash on hand. Crisis ensues. The country of Iceland is put up for sale. You serve ten months in minimum security, and upon release you're appointed Secretary of the Treasury.

True footnote: According to the CIA's World Factbook (which, government eavesdroppers take note, I found online) the combined national budgets of the 221 highest-spending countries, from the United States to Zimbabwe, is currently $24 trillion. In late 2007, leading up to the financial crisis, the worldwide value of outstanding credit default swaps (money owed by the banking industry in case of default) was over $62 trillion. If you were wondering what could possibly have panicked world leaders into cooperation, that was it.

The illusion of skill is one way we misjudge reality in our favor. Research has shown that we believe we're better than we are and that we'll be more successful than we will. Gyms, diet companies, and dating sites trust us to overestimate our chance of success, even if we're aware of the relevant statistics. I love weddings—the optimism is palpably overwhelming—the only people not smiling in bliss are the ones crying with joy. It's false optimism of course, as everyone knows, but through our rose-infused glasses it's easy to see that the stats won't affect *us*. Stanford University doctor Richard L. Peterson claims, based on fMRI studies of investors, that overconfidence has a stimulating effect on the brain's dopamine reward system; it just feels good to be sanguine.

Psychologists note that while we unrealistically inflate our own merits, we tend to be accurate in assessing others'. Nicholas Epley and Erin Whitchurch performed a study in which they showed participants pictures of themselves, one an original snapshot and the others digitally altered to make the subject look less or more attractive. Both the subject and a stranger who had met with the subject a week earlier were asked to select the unmodified photo, and whereas strangers most often picked the original, subjects gravitated to images that made them appear better. And if you're thinking, "I myself wouldn't fall into a self-serving delusion like that," you just fell into one. We can see the speck in our brother's eye, but not the log in our own.

## LIFE IS A BEAUTY CONTEST

### Pop Quiz

What point guard scored the final point, a free throw, to win the 1982 state championship game, and two years later won her city's beauty pageant, where she was also crowned Miss Congeniality, then went on to be runner-up in the state pageant, and placed second again, with Senator John McCain, in the 2008 U.S. presidential election? Correct—Sandra Bullock.

Life is a beauty contest. Ask the "smart, sweet, and hard working" television character Ugly Betty (America Ferrera). "The only reason the publisher hired her to be his son's secretary is that he thought Betty was someone who Daniel would never sleep with." Or ask C.D. Bales, Steve Martin's character in *Roxanne*, a 1987 film version of *Cyrano de Bergerac*, who like Cyrano enlists a handsomer man to woo the woman he loves. "You know, you could de-emphasize your nose if you wore something larger, like Wyoming." Or ask a lawyer. In 2009, a disabled law student sued Abercrombie & Fitch for discrimination, claiming the retailer made her work in a stockroom because her prosthetic arm didn't fit its public image.[94] Or ask Aristotle: "Personal beauty is a greater recommendation than any letter of introduction." Or if you want the plain truth then ask a scientist, because the evidence is overwhelming and incontrovertible. Life is a beauty contest.

Attractiveness bias is our innate tendency to judge good-looking people positively, even in situations where appearance is irrelevant. And although you and I would never succumb to "lookism," humankind's preference for pulchritude has reared its opposite-of-ugly head in every nook and cranny that researchers have peered into. Lookism affects how teachers judge students, how voters choose candidates, and how juries decide cases. Psychologists Stephen M. Smith, William D. McIntosh, and Doris G. Bazzini reviewed five decades of top-grossing U.S. films, and found that attractive characters were consistently portrayed more favorably than unattractive characters.[95] In employment decisions, looks influence both hiring and firing. Research reveals that juries are more likely to acquit attractive defendants, and that judges typically give them lighter sentences if convicted. (Haidt's take is that the elephant tips off the rider to interpret evidence from an angle that supports the elephant's inherent desire to acquit.) And in case you're wondering whether beauty might be cultural, even babies express a preference for attractive faces.

Life is a beauty contest in more ways than one. Our evolutionary urges reward not only physical allure, but cleverness and dominance as well, because all three signal reproductive fitness. Looks, brains, and might—I possess them of course—but on what grounds can I claim them as mine? For better or worse, and it depends on bestowed attributes, life is a beauty contest where some fortuitously win and some unluckily lose.

Miss Congeniality managed to accomplish both by raking in millions while remaining mired in her "Don't retreat, reload!" mentality.

## CALL ME PATSY

I'm a sucker. Marketers know more about my inborn appetites than I do. Psychologist Robert Cialdini's book *Influence*, currently in its fifth edition, describes how "compliance professionals" manipulate our automatic responses.

Did I mention I'm lazy? Well I'm absolutely too lazy to scribble my return address at the top of an envelope, so it's lucky that charitable organizations keep sending me those little labels. Am I a bad man for not subsequently donating? They hope I'll feel that way, and research suggests that I will, because humans are innately primed for reciprocity. Negotiators will sometimes invert the reciprocation ploy by making the first "concession."

Sales agent: "Would you buy the car if I could get your price?" Say yes and you've admitted that you want the car, say no and you seem foolish for snubbing a potentially great deal, name a lowball figure and you appear unscrupulous, plus you're already negotiating price. People inherently want to be and to appear consistent; we like to follow up on what we say. Compliance professionals bait us into making an initial commitment because we're then more likely to carry through with it.

Toy sales can be lean after Christmas, so some companies have adopted a new strategy. They run ads prior to Christmas for toys they expect to be popular, then understock the toys. They know that some parents will have promised their child a specific toy, and that when they can't find the toy they'll buy a different one. After Christmas, the toy companies again promote the original items, kids become reenergized, and parents fulfill their earlier promise.

Mom: "If your friends jumped off the Brooklyn Bridge would you jump too?" Probably. Nature favors those who fit in with others, which is why marketers use social proof (people like us or people we want to be like) to sell everything from bourbon to fine art. When you watch a TV sitcom do you like the canned laughter? Nobody does. But producers use it anyway because studies have shown that it works. We're naturally inclined to laugh along.

Evolution has endowed us with an automatic and understandable respect for power, which is why authority figures and experts are always trying to sell us something. To defend against the appeal of entertainers, doctors, athletes, and other influence peddlers, Cialdini suggests we ask two questions.  First, do they possess relevant expertise, and second, how truthful are they likely to be?  I love Catherine Zeta-Jones, but do I really want to take her advice about cell phone plans?

Marketers intimidate us with the scarcity pitch.  It's not difficult to grasp why we're programmed to respond instinctively to insufficiency, and if you don't believe me, remember, this is a limited time offer, only two hundred will be sold at this price, and I have other people interested in the property.

## WE NEED TO TALK

> ### We Should Give Ourselves a Good Talking-To
>
> John I:I:  In the beginning was the Word, and the Word was with God, and the Word was God.  Case closed.  Language is paramount, possibly Almighty.  What better tool does God have with which to inject Himself into a nattering human mind?  The Mormon Tabernacle Choir?  John is a bit enigmatic, though, in that he suggests you can be with a thing at the same time that you happen to be that thing, but we'll let the paradox pass for now unless you'd like a moment with yourself to reflect on it.

Nobody knows how language originated.  Alfred Russell Wallace, best known for independently conceiving the theory of evolution via natural selection (he and Darwin co-published a paper on the subject in 1858), exempted from his evolutionary theory the human capacity for language.  He believed that the mechanism was too complicated to have evolved through natural selection, and must have arisen thanks to divine intervention.  Famed linguist Noam Chomsky too believed that language was too complex and sophisticated to have developed via the random progression of natural selection, though he allowed that when you cram billions of neurons into a tiny skull new laws of physics might emerge.[96]  Among divine intervention, new physical law, and natural selection, most scientists favor the last.

Scientists estimate that language originated some time between several million and forty thousand years ago, depending on where they draw lines between pre-language, proto-language, and language.  Steven Pinker refused to speculate about the feral precursors of language, but he vowed that the long transition from animals squawking to humans talking must have been gradual.  We might say that in the evolution of language a growl became a vowel, a screech a part of speech, a quack and a bray hackneyed word play. But no matter what we say, or how badly we say it, we should never underestimate the disparity between human language and animal communication, or the critical role of language in humanity's ascendance. It's fun to text using our opposable thumbs, and standing upright is handy for reaching the liquor cabinet, but language is what makes us human (or in the case of rhyming writers, subhuman).

But what does the fairly arbitrary set of symbols we call language actually do for us? First, it's the prime apparatus used in thought and symbolic reasoning. Many scientists believe that abstract thinking, the ability to reason with concepts rather than objects, is what most sets us apart from other species. The second great advantage of language is that we can chronicle and pass along what we've learned. Most human knowledge has now been recorded, and nearly all of it (in addition to countless inane opinions) is available for the googling. Third, thanks to our cerebral web of symbols and concepts, and the ability to manipulate them, we're able to work through contingencies in our minds. We can momentarily sidestep the hazards and inconvenient logistics of physical reality; we can make plans, run them on our internal simulator, and forecast results. We can, and we do, maintain an ongoing mental model of the world, a simulation that includes and features our selves.

---

### The Power of Words

Right now, as you read these words and process these concepts, I am literally carving neural pathways in your brain. I'm shaping in some small way who you are, which I hope you and your attorneys won't hold against me later. And of course, via language, you too have the power to alter minds. For instance, next time you're in a pub where some people are sharing stock tips, you can walk them over to the dartboard and say, "this is all you really need." You'll have assimilated the ideas in this book, which I've clearly liberated from others, refashioned them in your own style, and passed them on. What could be more human than that?

---

When Nobel laureate physicist Richard Feynman was twelve years old he taught himself to fix radios. His approach, often successful, was to open the radio, silently think through the possibilities using his basic knowledge of electricity and mechanics, and then enact a repair. Word of mouth spread that "he fixes radios by thinking."[97] We needn't be as clever as Feynman to recognize the dramatic evolutionary advantage of a mechanism, the mind, that can generate a manipulable model of reality.

A computer can run models, but mind is superior because we don't have to lug the thing around, program it, or deal with Windows. Instead, mind comes built-in, and is partly, largely, or entirely preprogrammed depending on which scientists and philosophers we listen to.

For those of us who enjoy being at the top of the food chain, language and symbolic reasoning are true blessings, yet the unshakable omnipresence of words, thoughts, and concepts obscures an underlying irony, which is that symbols and abstract representations of reality, by definition, aren't reality. Natural selection, of course, doesn't care about the distinction, because if living inside our remarkable minds somehow empowers us to better pass along our genes, then that's where we'll find ourselves—lost in thought. There's a downside to our bustling minds, but it's the subject of another chapter.

## MIRROR MIRROR

| Yada Yada Yada |
| :---: |

Elaine, talking about her shrink during an episode of Seinfeld:

*(Actual Dialogue)*

| | |
| --- | --- |
| Elaine: | He has this power over me, okay. He has this way of manipulating every little word I say. He's like a Svenjolly. |
| George: | Svengali. |
| Elaine: | What did I say? |
| Jerry: | Svenjolly. |
| Elaine: | Svenjolly? I did not say Svenjolly. |
| Jerry: | George? |
| George: | Svenjolly. |
| Elaine: | I don't see how I could've said Svenjolly. |
| Jerry: | Well, maybe he's got, like, a *cheerful* mental hold on you.[98] |

For Svengali, mind-control meant first hypnotizing a woman and then priming her with suggestions. He confessed exactly that to a female conquest in the 1934 film version of his story. "You are beautiful, my manufactured love. But it is only Svengali talking to himself again." I can outdo Svengali because I'm not a fictional character, I'm a scientist, and using my bona fide modern-day expertise I can sway a person's mind without Svengali's hocus-pocus, via mirror neurons.

In the early 1990s, Italian researchers discovered that a particular set of neurons would fire in a monkey's premotor cortex whenever the monkey reached for a piece of food. They then noticed that when the monkey saw a human reach for a piece of food, which in Italy is quite often, the identical neurons in the monkey's cortex would fire. The monkey's brain mirrored the researcher's action. Mirror neurons are now regarded as one of the most important neurological discoveries of recent times, but when Rizzolatti, Di Pellegrino, Fadiga, Fogassi, and Gallese submitted their findings to the scientific journal *Nature*, the manuscript was turned down because, "we don't publish articles about pasta, and besides, nobody in America is going to believe your names." Actually, it was rejected "for lack of general interest."[99]

From an evolutionary perspective, mirror neurons help us learn quickly via imitation. Scientists also believe that mirror neurons may contribute to the human ability called Theory of Mind, in which we can construe what's going on in another person's head, a handy skill in social situations.

Mirror neurons fire automatically in response to other people's actions, which is one reason movies and theater are popular—we really can experience other people's lives—and it's part of the reason that the laugh tracks used on sitcoms work, even though everybody hates them. In other words, Bill Clinton was telling the truth, at least this once, when he told us, "I feel your pain." Kirk, Spock, and McCoy (from *Star Trek*, for those of you who just woke up from a 60-year coma) regularly encountered "empaths," psychic creatures from whom they had to shield their minds. We're one of those species. As Giacomo Rizzolatti, head of the team that discovered mirror neurons, put it, "They [mirror neurons] allow us to grasp the minds of others not through conceptual reasoning but through direct simulation; by feeling, not by thinking."[100] Brain imaging studies have revealed that people who suffer from "mirror-touch synesthesia," in which their mirroring systems are overactive, literally find it difficult to distinguish between what's happening to them and what's happening to someone else.[101]

---

*How to Win Friends and Influence People*

To get on someone's good side, here's a proven tip: subtly emulate their body postures (not making this up). Dutch researchers recently found that mimicry might have adaptive value beyond the acceleration of learning. They suggest that physical imitation may "immediately and directly enhance the prospects of successful procreation for individuals who adopt the behavior." Ordinarily, a mind-control technique this potent should only be attempted by a professional researcher on a closed campus, but if you must try it on a soon-to-be "friend," at least be cheerful. What kind of results can you expect? On my upcoming trip to Los Angeles I'm having dinner with Jennifer Lawrence. She just doesn't know it yet.

## MOTIVATED IS MY MIDDLE NAME

### None So Blind as Me

It's easy for me to see that Taylor Swift is kidding herself when she says, "I'm the girl who rarely has a boyfriend," because according to professional matchmaker Patti Stanger, "Taylor Swift dates guys so she can write a breakup song about them."[102] Now, I've noticed that Swift writes a lot of breakup songs, but what's hard for me to see is that despite her self-proclaimed availability, and for a dozen obvious reasons, I have no chance with her.

Richard Feynman cautioned scientists against tainted objectivity. "The first principle is that you must not fool yourself, and you are the easiest person to fool." It's easy to dupe ourselves because mind is a master conjuror. Benjamin Franklin noted that, "So convenient a thing is it to be a reasonable creature, since it enables one to find or make a reason for every thing one has a mind to do." Scientists refer to our self-serving inner accounting system as motivated reasoning, and studies have revealed the wealth of tricks we use to reach the conclusions we'd like to reach.

Let's first consider the neurology of choice. In fMRI studies that looked into the neural basis of decision-making, researchers have been consistently able to predict a subject's decision before he or she becomes aware of it. In one study, where subjects were asked to press a button with their left or right hand, investigators could predict the chosen hand up to seven seconds before subjects pressed the button. "Your decisions are strongly prepared by brain activity. By the time consciousness kicks in, most of the work has already been done," said John-Dylan Haynes, a neuroscientist at Max Planck Institute who coauthored the enquiry.[103] In another study, researchers assessed brain images as subjects played the *Ultimatum Game*, a negotiating matchup in which two players decide how to divide a sum of money. The scientists could forecast with 70%

accuracy how players would proceed, and could discriminate between brain processes related to emotion and those that involved the objective evaluation of financial incentives.[104] Researchers in a comparable study concluded that, "the outcome of free decisions can be decoded from brain activity several seconds before reaching conscious awareness."[105]

The point is that before System 2 kicks in, System 1, operating outside consciousness, has already reached a verdict. What studies in motivated reasoning demonstrate is that the judgment is nearly always in our favor. After reviewing the vast literature on motivated reasoning, cognitive scientists Hugo Mercier and Dan Sperber concluded that the "depressing" findings make perfect sense once we understand that reason evolved not primarily to find truth, but to help us argue with, persuade, and manipulate other people. As Mercier and Sperber point out, skilled arguers aren't after the truth, they're after arguments that support their views.[106] This would help explain why confirmation bias is so powerful and enduring, and why Fox News and MSNBC almost always outperform CNN. On the basis of fMRI studies, scientists suggest that it's because our pleasure centers are activated when we "prevail."

Researcher David Perkins found that IQ was the biggest predictor of how well people argued, but it predicted only the number of my-side arguments. He concluded that people invest their IQ in buttressing their own case rather than in exploring the matter more thoroughly and evenhandedly.[107] For Jonathan Haidt, the rider acts as a lawyer rather than a scientist, and its job is not to seek truth, but to advocate for the elephant in the court of self and public opinion. The rider, rationality, acts as spokesperson for the elephant even though it may not know what the elephant is thinking. Put differently, the elephant carries a full-time public relations firm on its back, and it's in the employ of the elephant.

---

### Are Corporations People?

It's a red-hot question, but I ask you more importantly, are people corporations? We have in-house lawyers, in-house PR, and a product to sell (moi). While some of us are always looking for investors, others are relentlessly nonprofit. Our self-worth is a publicly-traded commodity, plus we answer to shareholders (family and pets) as well as a Board of Directors (family and pets). We advertise, compete, and eventually go out of business. We engage in horizontal and vertical mergers, though most people prefer to merge horizontally. Our decisions are motivated by profit.

Research indicates that instead of forming judgments from the bottom up, using data to draw a conclusion, we more typically decide top-down, using our preferred conclusions to shape the assessment of data. Political scientist Howard Margolis states: "Given the judgments (themselves produced by the nonconscious cognitive machinery in the brain, sometimes correctly, sometimes not so), human beings produce rationales they believe account for their judgments. But the rationales are only ex post rationalizations."[108]

Confabulation is a term psychologists use to describe our innate ability to fabricate post hoc justifications for our behavior. Psychologist Michael Gazzaniga, one of the first to study confabulation, dubs System 2 our interpreter module. The interpreter module maintains a running commentary on whatever the self is doing, even though it may have no access to the real causes or motives behind the self's behavior, and as Gazzaniga's work shows, the interpreter module is skilled at contriving explanations without realizing it has done so. The interpreter module is not unlike the narrating self, in that its evolutionarily-induced calling is to spin out a commendatory life story.

## MEMORIES OF ARNOLD

### The Life You Think You've Lived

Philip K. Dick's science fiction story "We Can Remember It for You Wholesale" is about memory implants, and even if you haven't read Dick's piece you're probably familiar with the story because it has twice (in 1990 and 2012) been adapted for films named *Total Recall*. The dialogue below, from the 2012 version, features Colin Farrell and Kate Beckinsale.

*(Actual Dialogue)*

Doug Quaid:   Tell me what's going on. Talk! Or we skip to "Until death do us part."
Lori Quaid:   I'm not your wife.
Doug Quaid:   That's bullshit. We've been married for seven years.
Lori Quaid:   I'm U.F.B. police intel, assigned to play your wife. Six weeks ago I didn't even know you.
Doug Quaid:   What are you talking about?
Lori Quaid:   It's true. Your memory was erased, your

> mind was implanted with a life you think
> you've lived. You keeping up, baby? There
> is no Douglas Quaid, there never was.
>
> Doug Quaid:  Are you saying I don't ... this ... every ...
>               our marriage . . .
> Lori Quaid:  What can I say? I give good wife.
>
> Hear me now and believe me later: Your own mind, too, has been
> implanted with a life you think you've lived. You keeping up, baby?
> There is no [fill in your name here], there never was. What can I say?
> You give good self.

Memory implants sound far-fetched because we believe that our memories are unblemished recordings and that we own them, but false memories aren't science fiction, they're another one of mind's mirages. Psychologist William L. Randall describes memory as a compost heap, where recent events retain detail but soon break down and mix in, and though exceptional events take longer to decompose, the rest eventually become mush.[109] Studies by Elizabeth Loftus, a cognitive psychologist who became famous for her work on the fallibility of memory, indicate that memory is more like a Wikipedia page than a video recording. It's a transcription of history shaped by multiple people's perceptions and by constantly varying suppositions. Loftus has testified in (or consulted on) hundreds of trials involving eyewitness testimony, including those of the Hillside Strangler, Michael Jackson, Martha Stewart, Oliver North, and Phil Spector.

DNA tests have revealed that hundreds of innocent people were wrongly convicted due to eyewitness testimony. Loftus demonstrated that third parties such as police, lawyers, and medical personnel can introduce false facts into a witness's memory via leading questions and by suggesting their own point of view. Lawyers now closely question every witness regarding the accuracy of their memories, and about any possible "assistance" from other people. Still, built-in biases degrade memory's objectivity because we inherently want to cast ourselves in a good light, and because we want to appear consistent. On top of that, memory is recast each time we tell a story because we tailor the account for our audience. We come to believe our own fish stories.

Psychologist Harry P. Bahrick and his colleagues asked ninety-nine college freshmen and sophomores to recall the grades they received in high school. The students knew their recollections would be checked against transcripts, so there was no incentive to lie. Overall, participants remembered 70% of their 3,220 grades correctly. Their accuracy for A grades was 89%, B grades 64%, C grades 51%, and D grades 29%. So if you're depressed over a bad evaluation, cheer up—your rating will get better with time.[110]

Following a series of high profile child abuse cases in the 1980s, Elizabeth Loftus and others exposed a chilling example of implanted memories. What the researchers discovered, via numerous experiments, is that preschoolers, and to a lesser degree elementary school children, could easily be led to believe they'd been abused by an adult. What it took was leading questions, the reinforcement of particular answers, and repetition, exactly the techniques that therapists and police used to build a case against the children's "abusers." Eventually, most of the cases were dropped or overturned, and in the late 1990s several clinicians were successfully sued for implanting false memories of childhood sexual abuse, incest, and satanic ritual.[111]

Might it be possible to implant gratifying memories as well? In the first *Total Recall* Doug Quaid (Arnold Schwarzenegger) had a fling with Sharon Stone and then Rachel Ticotin, and in the 2012 version Doug (Colin Farrell) did likewise with Kate Beckinsale and Jessica Biel. Some people—I won't mention names—might pay good money for those memories.

### SECRET AGENT

The Shadow is a fictitious crime-fighting vigilante who "knows what evil lurks in the hearts of men." He can also "cloud men's minds," a power that I hope you by now understand is somewhat superfluous, but in case you're not convinced, here's a final example of how mind conjures up reality.

Agency detection is our tendency to ascribe events to a doer. It's the innate inclination to presume intentional involvement by a sentient agent in situations that may or may not entail agency. The disposition to perceive agents in our environment is not unlike our predilection for patterns, because in each instance the blind brush of natural selection has groomed us to err on the side of caution. The failure to recognize a pattern, or to detect an agent such as a predator or an enemy, could be deadly. Oxford psychologist Justin Barrett is among those who argue that human beings have an inbuilt Hyperactive Agent Detection Device, a cognitive module that readily ascribes events in the environment to the behavior of agents.[112]

Humans regularly over-attribute agency to animals (Lassie must be trying to tell us something), to machines (Microsoft Windows hates me), and to random events (Zeus is angry). Religion, naturally, disfavors the hypothesis that we reflexively envision agency. I won't belabor the point, but I'll cite the final two sentences from a lengthy article on the Christian Research Institute website, a piece by Paul Copan entitled "Does Religion Originate in the Brain?" Copan equitably presents the scientific case, reviewing evidence of our innate penchant for religion, then concludes, "Naturalistic explanations that suggest that theology is a useful fiction, or worse, a harmful delusion, fall short of telling us why the religious impulse is so deeply imbedded. If God exists, however, we have an excellent reason for why religion should exist."[113]

It's tempting to roll one's eyes at Copan's dialectic U-turn, first explicating and then in one sentence dismissing science, and at his final argument, but it's more important to consider an article by Sindya N. Bhanoo that showed up in the Science section of *The New York Times*: "Male Antelopes Scare Partners Into Sex."

During mating season a male topi antelope, to retain a female in his territory, will pretend to sense a predator. He'll run in front of her, freeze in place, stare in the direction she's going, and snort loudly, all of which is exactly what he'd do if he sensed a predator, except this time he's faking it. The female antelope generally retreats back into the male's territory, where he promptly mates with her. Bhanoo concluded that: "Humans are not the only mammals who have hyperactive agency detection devices that operate in accord with a cost-benefit calculus, or who deceive others to get what they want."[114]

### THE INANITY DEFENSE

Ladies and gentlemen of the jury, I submit that the defendant, moi, is not guilty by reason of inanity. If I didn't understand the utter muddle-headedness of that thing I just said or did, then I clearly cannot be held accountable. On the other hand, if I realized its foolhardiness but proceeded anyway then I can't control myself, and once again I'm not responsible.

Larry Winget, the self-declared Pitbull of Personal Development, lays out a hazily similar case in his book *People Are Idiots and I Can Prove It*. Winget presents evidence regarding health, finances, consumerism, business, and a category he calls "general idiotic behavior." For Winget, health idiocy includes smoking, eating fast food, and not exercising, while financial lunacy is on display thanks to credit card spending and the failure to save money. On the consumer front, we purchase products that promise weight loss without dieting or exercise, age loss via a cream, and the last razor we'll ever need (though we can get a second one for half price). Under general idiotic behavior, Winget includes playing the lottery, drinking and driving, destroying the environment, as well as believing in witches, psychics, and that Elvis is alive. Is it that we don't grasp how preposterous and potentially self-destructive these behaviors are, or are we simply out of control?

The actual insanity defense is based on a precedent established in 1843 after the attempted assassination of British Prime Minister Robert Peel by Daniel M'Naghten.

> At the time of committing the act, the party accused was not labouring under such a defect of reason, from disease of the mind, as not to know the nature and quality of the act he was doing, or if he did know it, that he did not know he was doing what was wrong.[115]

Our minds, largely or entirely, in sickness or in health, till death do us part, arise from our brains. The brain operates via neural circuitry that was laid out by genes then modified through experience. Which part of that are we responsible for? For genes it's close to zero percent, and for childhood events maybe five or ten percent tops. Adolescence and adulthood convey increasing accountability, unless we're not in our right mind.

On August 1, 1966, a heavily armed Charles Whitman climbed the tower building at the University of Texas in Austin and began sniping. Before he was shot by police, Whitman had killed 14 people and injured 32 others. Earlier that day he had murdered his wife and his mother.

When we hear a story like that we immediately ask, "Why?" How could someone do such a horrific thing? In the case of Charles Whitman there was never a definitive answer, but there was evidence. Whitman had left a suicide note on July 31, plus several handwritten and typed messages that he composed during the early morning of August 1. In the suicide note he wrote, "I do not really understand myself these days. I am supposed to be an average reasonable and intelligent young man. However, lately (I cannot recall when it started) I have been a victim of many unusual and irrational thoughts."[116] In one note he asked that when it was over an autopsy be performed to determine if there was a physical reason for his actions. He requested that any money remaining from his insurance, after paying his debts, be "donated anonymously to a mental health foundation. Maybe research can prevent further tragedies of this type."[17]

An autopsy was performed on August 2. The autopsy discovered a glioblastoma (a highly aggressive and invariably fatal brain tumor) in his hypothalamus. Whitman would not have lived out the year. On August 5, the Connally Commission, set up by Texas governor John Connally, concluded that the lesion, "conceivably could have contributed to his inability to control his emotions and actions."[18] Forensic investigators and neurologists subsequently supported this supposition, noting that Whitman's actions may also have been influenced by a difficult and sometimes abusive childhood.

For me, the Texas tragedy presents three morals: a) our minds, for better or worse, are byproducts of the brain, b) it's difficult to assess someone's accountability when contributing factors are beyond his or her control, and c) though challenging, it might prove more productive if we could dispassionately identify and constrain dangerous individuals instead of blindly hating them.

But that's not my main point. The point might be that because an unfortunate childhood and a brain tumor are basically bad luck, there but for the grace of God go I, but that's not quite it either. We know from family evidence that nurture was not on Whitman's side, but what about nature? The autopsy noted that there was a vascular deformation surrounding Whitman's tumor. The Connally Commission report stated that the deformation may have been congenital, predisposing Whitman to develop a glioblastoma. If so, and if one can accept that the tumor could have contributed to his state of mind and his actions, then maybe genes were not on Whitman's side either. But that's also not my main point.

What I'd most like to suggest is that soundness of mind is not a two-valued variable; it's not sanity versus insanity, health versus illness. Each of us does not stand, or perhaps sit nervously tapping a foot, on one side of the partition or the other. As I see it, Charles Whitman happened to land at one tail of the sanity distribution. At the other tail of the same bell-shaped continuum are those rare individuals who seem to genuinely have it together. Most of us find ourselves somewhere in the middle, but wherever each of us falls on the sensibility curve, the brains that landed us there were largely fashioned by nature and nurture, both of which are beyond our control. Some brain states may feel better and result in more positive outcomes than others, but the metric we call "mental health," to me, is impostorous.

The easiest way to understand what I mean is to break the term impostorous into its component parts: *impostor / o / us*. An impostor is a fraud, "o" is an exclamation of awareness, and us means society. I'm saying that within American culture soundness of mind masquerades as an attainment, when it might be better understood as the mental health segment of life's beauty contest. If I by chance score high on saneness, and especially if I win the swimsuit competition, how can I take pride in conferred attributes?

An actual heiress once said to me, "I'm proud to find myself in this position." I didn't call her on it because I liked her, and because we all occasionally issue myopic proclamations. Plus, since personality is fundamentally a product of factors beyond our control, and since her words represented nothing more than a momentarily inelegant mental formulation (technically known as a brain cramp), I had no choice but to find her innocent by reason of inanity.

Neuroscientist James Fallon studies the brains of serial killers. In 2005 he discovered that his own brain scan was a dead-on match for the typical psychopath's, prompting further tests that showed he was indeed aggressive, with little empathy for others. His "kills," fortunately, were restricted to games and debate, where he in fact showed no mercy or remorse. "I'm kind of an asshole," he admitted in *The Smithsonian*, "and I do jerky things that piss people off."[119] Buddy, welcome to the club.

In 2013, actor Alan Alda hosted a two-part PBS television series called *Brains On Trial*. "I was surprised to see how well brain scientists are beginning to put together what's going on inside our heads, some-times before we're even aware of what's going on in there ourselves," Alda said. "As I talked with scientists and jurists on this show, I became convinced that before this new research makes its way into the courts, we need to think about what it could mean to our system of justice." Tom Wolfe, author of *The Bonfire of the Vanities*, is pretty sure what it means. He smells a rat when neuroscience insinuates that, "The fix is in. We're all hardwired. You can't blame me. I'm wired wrong!"[120]

So are we programmed or free? Are we the marionettes or the puppeteers? Is life universal or personal? Should we get over ourselves or get into ourselves? Will I ever shut up or just keep blabbing? Stay tuned.

# EMOTIONS

*"We organize our lives to maximize the experience of positive emotions and minimize the experience of negative emotions."*
- Paul Ekman, psychologist

*"Shopping, orgasm, learning, highly caloric foods, gambling, prayer, dancing 'til you drop, and playing on the Internet: They all evoke neural signals that converge on a small group of interconnected brain areas called the medial forebrain pleasure circuit."*
- David J. Linden, neuroscientist

*"A woman in love can't be reasonable—or she wouldn't be in love."*
- Mae West, actress and philosopher

---

## *GOT THESE FEELINGS COMING OVER ME*

Do you control your thoughts?  Here's a brief test.  Close your eyes and for the next thirty seconds stop thinking.  Waiting ... waiting.  OK, so maybe you're not a yogi.  At least try this one:  Close your eyes and for ten seconds don't think of a pink elephant.  Waiting ... waiting.  Good attempt at a nice try—a pink elephant sighting sans tequila.  The fact is that thoughts simply bubble up into consciousness, and as brain imaging studies show, that includes thoughts as seemingly purposeful as decisions.  Still, a few of us will insist that some superordinate "I"—let's call it moi—orchestrates our thinking.

But nobody (except method actors and politicians) intentionally authors his or her emotions.  Feelings just come over us, sometimes as a gentle wave, sometimes like a tsunami.  They come and then they go.

Emotions are our aboriginal response system, a reactional mode that operates without thought, concepts, or deliberations.  Feelings arise instinctively in our mid and hindbrains then quickly infuse us, body and mind.  As soon as the elephant triggers an emotion, the rider, with limited insight, sets to work rationalizing its cause, whereupon the rider presumes it's back in control even though the cause is presumptive, and despite the fact that each emotion has a life of its own.

### Emotion Explained

The word "emotion" dates back to 1579, when it was adapted from the French word émouvoir, which means "to stir up."[121]  And if anybody's going to stir me up it might as well be Jennifer Lawrence, as in this scene with Bradley Cooper from *Silver Linings Playbook*.

*(Actual Dialogue)*

Tiffany:    Yes.  Do you feel that?  That's emotion.
Pat:        I don't feel anything.
Tiffany:    Has anybody ever told you how Tommy died?
Pat:        No.
Tiffany:    We were married for three years and five days, and I loved him.  But for the last couple months, I just wasn't into sex at all.  It just felt like we were so different and I was depressed.  Some of that is just me, some of it was he wanted me to have kids and I have a hard enough time taking care of myself.  I don't think that makes me a criminal.  Anyway one night after dinner he drove to Victoria's Secret at King of Prussia Mall and got some lingerie to get something going.  And on the way back, he stopped on 76 to help a guy with a flat tire and he got hit by a car and killed. And the Victoria's Secret box was still in the front seat.  That's a feeling.

If you avow it Ms. Lawrence, I believe it.  Only psychologist Paul Ekman knows more about emotions than you, but he was unavailable to discuss them over dinner, so what do you say?

Paul Ekman has been dubbed "the best human lie detector in the world."[122]  Specialists use Ekman's Facial Action Coding System (FACS) to categorize and quantify emotional expressions.  Animators use FACS to more realistically portray emotions, and law enforcement agencies, including the TSA, use it to ascertain whether someone is telling the truth.  The television program *Lie to Me*, which aired from 2009 to 2011, is based on Paul Ekman's work, and Ekman served as the show's chief scientific advisor.

Ekman believes, as did Darwin, that emotions evolved via natural selection.  If that's true, emotions should be universal across cultures, which is exactly what Ekman found in Papua New Guinea, the United States, Japan, Brazil, Argentina, Indonesia, and the former Soviet Union.

It's also true that people born congenitally blind exhibit the same array of facial expressions as the sighted, an additional piece of evidence that emotions are innate rather than learned.

In *The Expression of the Emotions in Man and Animals,* Darwin argued that emotions, and the ways in which we express them, afford a survival advantage. Ekman concurs: "Emotions evolved to prepare us to deal quickly with the most vital events in our lives. We have automatic appraising mechanisms that are continually scanning the world around us, detecting when something important to our welfare, to our survival, is happening."[123]

Emotions are how nature informs us of what to pay attention to. Emotion kicks in when there's danger, opportunity, or in any situation that might be directly relevant to our reproductive prospects. Moreover, evolution programmed us to distinctively remember emotional episodes, which means we become increasingly attuned to potentially significant stimuli.

Our emotional sensors, which Ekman calls "autoappraisers," and which activate in response to situational catalysts, nonconsciously and relentlessly scan our environment. But feelings can crop up without a direct environmental trigger. We can conjure up an emotion via mental reflection, we can empathically feel someone else's emotions, and we can even fabricate a feeling by forming its associated facial expression.

---

## *Courtship is a Marketing Ritual*

Kinesics is the study of body language—facial expressions, gestures, and posture—a communication scheme that predates speech. Today's Lotharios and Lothariettes employ romantic overtures such as, "Do you believe in love at first sight, or do I need to walk by you again," and, "If I asked you for sex would your answer be the same as your answer to this question," but prior to verbal entreaties, men and women relied on body language to convey their fleshly intent. Our unconscious and automatic courtship signals (which I'm not making up) persist to this day.

Men, it turns out, are not subtle with body language, and they typically broadcast their message to any woman within range. If you've ever been to a party, a wedding, a juke joint, a concert, a dog park, a company picnic, or even a funeral, you've seen men scanning the crowd for talent. They usually stand erect with shoulders wide and stomach pulled in, head on a swivel. Common postures include the "cowboy stance," with thumbs in belt loops and fingers pointing down towards the crotch, and "hands in pockets," except for the thumbs which remain out and point down to the crotch.

Evolution fashioned women to be more subtle and more selective. Miss Piggy speaks for many women, no doubt, when she says, "Moi speaks body language fluently, although with a slight French accent." The most accurate barometer of a woman's interest is eye contact, and as a rule her degree of eye contact correlates with her receptiveness. No eye contact equals no interest, sorry. But if a guy passes the eye exam, then further signs of attraction include parting or moistening of the lips, leaning forward, canting her head to the side, dangling a shoe, twirling her hair, or fondling a cylindrical object such as a pen, hanging earring, or wine stem (thank you Dr. Freud).

These are all helpful tipoffs, but it's more valuable to understand romantic signage within its greater context—marketing. It's best to recognize that courtship follows the AIDA advertising model (which I'm again not making up): Attention, Interest, Desire, Action. The first step, a sine qua non, is to attract the other's Attention, which is one reason that tattoos, body piercings, muscle shirts, plunging necklines, and Rolexes are popular. The attention stage is most critical in crowded, competitive venues such as nightclubs. As soon as the other becomes attentive, one needs to arouse Interest by proffering one's advantages and benefits, which can include intelligence, prosperousness, talent, physical prowess, kindliness, or plain old sexuality. Next, one must edge the prospect from Interest to Desire, which means getting him or her to commit, thus closing the deal. Only then comes Action, in the case of courtship rather literally.

Courtship is a marketing ritual. The AIDA model was introduced in the early 1900s, then quickly followed by the AIDAS model, which appended a Satisfaction phase to the end of the protocol. And don't we all want to leave our customers satisfied? (If you judge by my spam folder, there's nothing more important.) In 2011, business scientist Bambang Wijaya further advanced the marketing/courtship analogy when he introduced an advertising model called "AISDALSLove," in which he envisioned a "Pyramid of Love" built upon customer brand loyalty. And wouldn't we all like romantic patrons to be faithful to our personal product line? (Which, if we play our cards right, might at some point even come to include a non-metaphorical version of the pyramid of love.)

While emotions are common across cultures, the triggers that set them off need not be. There are elemental triggers, like fear of an object hurtling towards us, cultural triggers, such as disgust at public displays of affection, and individual triggers, like road rage, which engulf some of us but not others. Research shows that we have little control over our response to universal triggers, but that learned triggers can sometimes, with effort, be modified.

Emotion comes over us as a sequence of phenomena. First, an event or cue triggers an autoappraiser, which quickly and unconsciously classifies the circumstance as positive, negative, or immaterial. When the needle swings toward good or bad, the limbic system in our midbrain initiates an emotion, which primes our body and implants thoughts in our mind. What typically ensues is a self-reinforcing cycle of sentiment, bodily sensation, and thought, that strengthens our emotional state and propels us to act or stand down. Normally, our reflex emotional system works well, but it can misfire if the evoked emotion is inappropriate, is unfittingly intense or tepid, or is insensitively expressed.

Like all attributes, emotions vary from person to person, meaning that each of us exhibits an "emotional profile" for each type of feeling. We differ in how quickly the sentiment comes over us, how strongly it's felt, how long it lasts, and how much time it takes to recover.

Oprah keeps telling us our emotions are meant to be shared, and she's right—every individual has a built-in emotional signaling system, and the messages we send out are both involuntary and cross-cultural. When we broadcast anger, others back off. When we broadcast fear, others look around. When we broadcast sadness, others come to help. When we broadcast lust, any number of things could happen.

We were not designed to challenge our emotions. We've been wired instead to affirm them, which means that we evaluate whatever's happening in a way that's consistent with what we're feeling. We then rationalize our emotions by finding their "cause," almost always in some external circumstance, in spite of the fact that emotions are internally generated and are almost by definition irrational. Women react initially with intuition, and guys go with their gut, but in either case emotion leads the way.

## Men Have More Guts than Women

We men follow our gut. We follow it to hasty judgment, and we follow it, often at a distance, as we schlep through the doorway and across the room. Did you know that 95% of the body's serotonin is found in the bowels? Or that the bowel can go on functioning normally after its connection to the brain, the vagus nerve, has been severed? Or that there a hundred million neurons in the gut, more than in either the spinal cord or the peripheral nervous system? It's all true, and if you don't believe me read *The Second Brain* by Dr. Michael Gershon, Chair of the Department of Anatomy and Cell Biology at Columbia University Medical Center.

According to Gershon, humans have a master brain in the head plus an apprentice in the bowel, and while male and female crania are of nearly identical size, even a nonscientist can observe that men are superiorly gut-enabled. We men are proud to brandish our overabundant brainpower. In a *20/20* television interview with Barbara Walters, George W. Bush justifiably boasted, "I'm not a textbook player. I'm a gut player. I rely on my instincts." The President was backed up by Stephen Colbert, who spoke during Bush's 2006 White House Correspondents' Dinner: "That's where the truth lies, right down here in the gut. Do you know you have more nerve endings in your gut than you have in your head? You can look it up. Now, I know some of you are going to say, 'I did look it up, and that's not true.' That's 'cause you looked it up in a book. Next time, look it up in your gut. I did. My gut tells me that's how our nervous system works." Personally, I'm with Bush and Colbert; there's no evidence that men are brainier than women, we just know it in our amply endowed guts.

Emotions are not truly ours. They come over us from somewhere in our primordial depths, by which I mean our mid and hindbrains. The fact that emotions evolved via natural selection indicates that they have, or at least had, survival value, but like our inbuilt cognitive biases they sometimes lead us awry.

Angina pectoris is a pain in the chest, sometimes severe, caused by congested coronary arteries. Surgeons in the 1950s regularly tied off damaged arteries, believing that healthy new channels would develop in adjacent heart muscle, and patients consistently reported a noticeable lessening of pain following the operation. It was only after a few patients had died that autopsy pathologists discovered there were no new blood vessels, which led befuddled surgeons to conduct, in 1958, an experiment that would never be allowed today. They performed the normal tie-off operation on thirteen patients and a sham operation on another five, where they simply cut through the skin and stitched it back. Eleven of the "real" patients reported relief, as did all five of the "sham" patients.[124] It was the placebo effect writ large. Feelings of cardiac stricture were faked out by a surgical sugar pill. If primal pain can be deluded, to what degree can we trust our less consequential feelings?

Like perceptions and memories, emotions are constructed based on incomplete nonconscious data. Autoappraisers respond heuristically to selected highlights of an event, relay their impressions to convergence centers that factor in preexisting biases, beliefs, and expectations, after which the mental-emotional payload is passed along to consciousness and into our bodies. That's the reason angina pain can improve with no change in the nerve impulses arriving from the chest. Because emotions are unconsciously engineered from imperfect data, we can end up with skewed sentiments as well as with outright emotional miscalculations, analogous to optical illusions and false memories.

Yet we instinctively embrace our feelings, project their cause onto the environment, and justify them at the expense of others. Even though emotions are beyond conscious control, once they inhabit us we take them to be ourselves and we act on their behalf. It's adverse possession, or on occasion demonic possession, from within. Maybe Oscar Wilde put the best spin on our situation. "The advantage of emotions is that they lead us astray."

## *FEAR*

### *Nothing to Fear*

We have nothing to fear but fear itself. That and getting blown to smithereens. Plus scorpions, flesh-eating bacteria, sharks, tornadoes, sharknadoes, the national debt, and the blue screen of death. So forget about what FDR said. Be afraid. Be very afraid.

A good place to start is with the local news. Fear sells because nature directs us to pay attention to things that could harm us, hence, "New Study Shows Link Between Cell Phones and Cancer. Details at Ten," and, "Contaminated Fish Poisons Family of Three. Could You Be Next?" Security expert Gavin de Becker suggests we pass on the local news because it fosters pointless anxiety, provides few real safety tips, and misleads us about what's actually dangerous. On top of that, he points out that unwarranted fear is a form of victimization. Talking heads will warn us about the cell phone menace, but we'll probably never hear the following candid headline. "If You Smoke Cigarettes Don't Worry

About Your Cell Phone—Quit Smoking!"

And now from our investigative news team. "Don't Become the Next Victim of Baggage Stalkers." They hide in your luggage before it reaches the airport carousel, then rob you at home. Here's what to check for: Bags that seem extra heavy, noises coming from your luggage, and bag contents that move around. Here's how to protect yourself: Kick each suitcase before removing it from the carousel, place loaded mouse traps in your bags, and always carry a gun to the airport. New at ten— "A Stalker's Perspective—Our Video Crew Goes Inside Your Luggage."

In *A Natural History of the Senses*, author Diane Ackerman tells us, "The brain is a good stagehand. It gets on with its work while we're busy acting out our scenes. When we see an object, the whole peninsula of our senses wakes up to appraise the new sight. All the brain's shop-keepers consider it from their point of view, all the civil servants, all the accountants, all the students, all the farmers, and all the mechanics."[125] Gavin de Becker recommends we add "all the soldiers and guards" to the top of Ackerman's list, because their judgment overrides the rest.

Fear trumps other feelings. Animals will bolt from food or sex if they feel threatened. There's been more scientific research on fear than any other emotion because it's easy to arouse fear in an animal. Fear is characterized by the threat of physical (and in humans psychological) harm, and as with other emotions, the triggers that provoke fear can be universal, cultural, or personal. According to Ekman we can learn to be afraid of almost anything, and it requires sensitivity and compassion to respect and reassure someone who's afraid of something we're not afraid of.

When fear is aroused our first instinct is to freeze or flee. Video footage taken during the 1996 Atlanta Olympic bombings shows that the majority of bystanders froze when the bombs detonated. The next most likely response is to direct anger at the threat. Whichever happens, and only protracted training can modify an individual's natural reaction, our bodies prepare. Blood flow increases to our arms and legs (for fighting or running), our adrenal glands release cortisol (to help blood coagulate quickly in case of an injury), our vision narrows and focuses, lactic acid heats up our muscles, and our heart and lungs ramp up support for the entire effort.

One of Bill Clinton's twenty favorite books is the 1974 Pulitzer Prize-winning *Denial of Death* by cultural anthropologist Ernest Becker. (When I looked up Clinton's other nineteen favorites, one book that was not among them was Gaynor Arnold's *Girl in a Blue Dress*.) In *Denial of Death*, Becker states that, "animals, in order to survive, have had to be protected by fear responses," the consequence of which has been, "the emergence of man as we know him: a hyper-anxious animal who invents reasons for anxiety even when there are none."[126]

Fear is of looming danger, an intense immediate threat. It links directly to the imminent possibility of pain or death, but its emotional progeny, anxiety and worry, concern themselves instead with some sort of hypothetical future. Authentic fear is undeniable, but anxiety, the vague unease brought on not by events but by imagination or by nothing discernable, nonetheless evokes a nagging fear reflex.

I ask again, do we control our thoughts? I ask because worry is the fear we fabricate in our minds, even though most of us understand that worry is neither productive nor healthy. Perhaps we could we fall back on the inanity defense, the devil makes me do it, or is it possible that worry is somehow rewarding, that it might relieve us of the need to actually do something about the hypothetical "problem"? In his book *Emotional Intelligence*, psychologist Daniel Goleman contends that we utilize worry as a "magical amulet" to ward off danger. We believe that worrying about something might somehow prevent it from happening. Goleman says we frequently worry about low-probability contingencies, because we act on the possibilities that seem likely, which means that worry is often squandered on unlikely eventualities.[127] As Mark Twain confessed, "I have had a great many troubles, but most of them never happened."

### ANGER

For Paul Ekman, angry feelings can range from minor annoyance to rage. Indignation, for example, is self-righteous anger, while sulking is passive anger, and exasperation is patience pushed beyond the limit. Revenge, as Ekman notes, can be harsher than the act that provoked it. Prolonged resentment turns into a grudge, and when bad blood festers it can be difficult to get the object of ire out of mind. In Garrison Keillor's words, "That's what happens when you're angry at people. You make them part of your life."

Hatred is an intense enduring enmity. We're not continuously irate, but real or imagined encounters with the despised one(s) can easily reawaken anger. According to Ekman, hatred and enduring resentment aren't emotions because they last too long, and they're also not moods, which last longer than emotions but not a lifetime. Ekman's point is that while deep-rooted hatred and resentment arouse bitterness, they aren't elemental anger.

Primal anger is immediate and intense. It's the brutish heart of attack and violence, as well as the most dangerous emotion to others. From an evolutionary perspective, anger directs us to dispatch our rivals, physically or psychologically, and to defend ourselves. "Back off." "Get out of my way." "Don't touch the remote." Research has shown that the easiest way to enrage an infant is to restrain its arms—even babies are ready to battle when their intentions are thwarted.[128] Fortunately, most don't begin karate lessons until age five.

Anger inflames anger in others, and the reciprocating cycle can rapidly escalate. One of our instinctive responses to anger is retaliatory hostility, particularly when the other person's anger seems unjustified. Disappointment in another's behavior can also make us angry, especially if that person is someone we care about, and though it might seem odd to be angry with those we love, they're the ones who can most hurt and disappoint us. Gavin de Becker points out that a woman in the United States is killed by an intimate every two hours. "If a full jumbo jet crashed into a mountain killing everyone on board, and if that happened every month, month in and month out, the number of people killed still wouldn't equal the number of women murdered by their husbands and boyfriends each year."[129]

In *Anger: The Misunderstood Emotion*, psychologist Carol Tavris argues that getting your anger out, an approach that some advocate, usually makes matters worse. After reviewing the research, Tavris came to the conclusion that suppressed anger doesn't, in any predictable or consistent way, make us depressed, produce ulcers or hypertension, set us off on food binges, or give us heart attacks.[130]

---

### Anger Management

It's clear that some groups hate America. Within U.S. borders it's either liberals or conservatives depending on which television station you watch, and on the international scene there are any number of terrorist sects that have sworn to rain death on the United States. Their hatred stems from two factors: an overabundance of testosterone and a dearth of racquetball courts.

Philosopher Ken Wilber calls testosterone "the fuck it or kill it hormone," and with the sole exception of political commentators James Carville and Mary Matalin, and the possible exception of comedian Bill

Maher and bleating heart Ann Coulter, ideological squabbling rarely results in the former. If you want a better world elect women, because their testosterone level is one-tenth of a man's. Data show that female legislators worldwide, by and large, support children, family, community, cooperation, consensus-building, and simple tolerance. Perhaps it's just coincidence, but all nineteen of the 9/11 hijackers came from countries where women are second-class citizens. According to gender experts in a 2013 Reuters Foundation poll, even the most liberal of the terrorists' home nations, the United Arab Emirates, still allows men to physically discipline their wives.[131]

Why racquetball? Because exercise mitigates anger. In a recent study, stress physiologist Nathaniel Thom enlisted sixteen young men with "high-trait anxiety" (a short fuse), then fitted them with "high-tech hairnets" that could read electrical brain activity. He showed the men incendiary photographs depicting Ku Klux Klan rallies and children under fire from soldiers, and found that those who'd ridden a stationary bike for thirty minutes were less incensed by the photos. Thom concluded, "If you know that you're going to enter a situation that is likely to make you angry, go for a run first."[132]

That's right, just jog into the custody hearing wearing soppy stinky sweats and panting like a St. Bernard. It will re-enchant your ex and endear you to the judge. Luckily, there are other ways to curb hostility, and for those we turn to Dr. Buddy Rydell (Jack Nicholson) and Dave Buznik (Adam Sandler) in the film *Anger Management*.

*(Actual Dialogue)*

Dr. Buddy Rydell: Now then we need to go over some ground rules. You are to refrain from any acts of violence including verbal assault and vulgar hand gestures. You may not use rage enhancing substances, such as caffeine, nicotine, alcohol, crack cocaine, slippy-flippies, jelly stingers, trick sticks, bing bangs, or flying wil-lards.

Dave Buznik: How about fiddle-faddles?

Dr. Buddy Rydell: Under my supervision. Also, if you are unable to stop masturbating, please do so without the use of any pornographic

|                     | images depicting quote, unquote "angry sex." All that having been said, I'm a pretty good guy and I think you'll be pleasantly surprised how much fun we can have together. |
|---|---|
| Dave Buznik: | Geez, without slippy-flippies or angry masturbating, I don't see how that's possible. |
| Dr. Buddy Rydell: | Sarcasm is anger's ugly cousin, Dave. |

## HAPPINESS

People say there's more to life than money because money can't buy happiness, and happiness is what life's really about. They're wrong on two counts. First, money *can* buy happiness, and second, happiness is not what life's about. If your son purchases Microsoft's "Experience Machine" you'll understand what I mean. Microsoft stole the idea from philosopher Robert Nozick, because Microsoft expends all its creative capital figuring out where to locate the Windows Start button.

Anyway, your son brings the Experience Machine home, plugs it in, thumbs the hedonic slide to "True Happiness," dons the headpiece and promptly flops onto the couch. The first day he sofas out in bliss for two hours, the next day for five hours, and from then on for eight or nine hours a day. As a parent, you're not impressed. But what if I told you it's called the Experience Machine because rather than simply stimulating the brain's pleasure centers, it activates circuitry that fully evokes every activity, perceptually rich and physically satisfying. He lives through and savors each new undertaking. We know that the experience of external reality is brain-created to begin with, and now your son is engaged in dynamic, constructive endeavors. In short, money has bought your son genuine happiness, and he's got the brain images to prove it.

But given all that, we recognize that something is intrinsically wrong with this picture. Nonetheless, the apparatus becomes a bonanza for Microsoft until machines worldwide stop working. Apparently, the Start button has disappeared.

### Legal Disclaimer

On advice of counsel, I have up until now withheld the fact that Microsoft does not actually manufacture an Experience Machine. I made it up. Thanks to publisher Larry Flynt, evangelist Jerry Falwell, and the

United States Supreme Court, I'm free to fabricate that sort of facetious information. In *Hustler Magazine v. Falwell,* a 1988 case immortalized in Oliver Stone's movie *The People vs. Larry Flynt,* the Supreme Court confirmed that satire is protected under the First Amendment.

Briefly, *Hustler* had run a fake ad which implied that Falwell had had sex with his mother in an outhouse. The ad was a parody of the then-popular Campari (an Italian apéritif) advertisements featuring celebrity interviews. In his "interview," Falwell confessed that he and his mom were, "drunk off our God-fearing asses on Campari," that, "Mom looked better than a Baptist whore with a $100 donation," and that, "I always get sloshed before I go out to the pulpit. You don't think I could lay down all that bullshit sober, do you?"[133]

The Experience Machine is, in reality, a thought experiment by Robert Nozick designed to illustrate the complexity of human happiness. Gratification is simpler for rodents. If you insert an electrode into a rat's cerebral pleasure center, then connect the electrode to a button in its cage, the rat will push the bliss button, while ignoring the food button, until it starves. Rodents go for cheap thrills; would that it were so easy for us. Human beings, for reasons that predate the Puritans by many centuries (although we shouldn't diminish their contribution), tend to stigmatize artificial and unmerited pleasures like drugs and paid-for sex, and we often make them illegal. Humankind, it turns out, would rather get its kicks the old-fashioned way, by earning them. That's why Miles Monroe (Woody Allen in *Sleeper*) refused to step into the Orgasmatron. "Machine? I'm not getting into that thing. I'm strictly a hand operator, you know. I ... I ... I don't like anything with moving parts that aren't my own."[134]

Paul Ekman insists that the terms happiness and enjoyment are too general. He proffers, instead, a set of pleasant feelings that includes sensory pleasure, amusement, contentment, excitement, relief, gratitude, ecstasy, awe, and wonder. He notes that each of us favors one or more of the pleasurable sensations over others, and that most of us organize our lives to maximize our most sought-after positive feelings.

In spite of the fact that the Declaration of Independence reveres "the pursuit of happiness" as highly as life or liberty, and despite the fact that happiness is many people's greatest life motivator, mental health professionals have traditionally focused on its opposite—dysfunction. The field of "positive psychology" wasn't formally introduced until 1998, when Martin Seligman and Mihaly Csikszentmihalyi launched their new approach with the goal of making everyday life more fulfilling.

Originally, Seligman believed that happiness derived from three factors: positive emotion, engagement, and meaning. Positive emotion, for Seligman, refers to the set of feelings that Ekman identified, such as amusement, contentment, and gratitude. Engagement is the loss of self-consciousness (which Seligman and Csikszentmihalyi call "flow"), that occurs during engrossing activities. Meaning results from recognizing and trusting something bigger than oneself. Recently, Seligman added relationships and achievement to his list, because research has shown that each of the two contributes to moment-by-moment happiness and long-term life satisfaction.

Brainiac Level Pop Quiz: According to Barbara Fredrickson, the principal investigator with the Positive Emotions and Psychophysiology Lab at the University of North Carolina, who is the "queen of happiness"? Hint #1: It's not Oprah Winfrey. Hint #2: It's definitely not Sarah Palin. Hint #3: Her first name is Sonja, and her last name looks like a box of alphabet cereal exploded. Give up? According to Dr. Fredrickson, the queen of happiness is psychologist Sonja Lyubomirsky. (Amazing work if you got that one.)

Lyubomirsky's book, *The How of Happiness*, delivers on its title. If you want to know what makes people happy, and what you might do to enhance your own happiness, Lyubomirsky folds open the scientific recipe. Harvard psychologist Daniel Gilbert states bluntly that unlike most books about happiness, Lyubomirsky's "happens to be true."[135]

Lyubomirsky tells us that: "Whether you drive to work in a Lexus hybrid or a battered truck, whether you are young or old, or have had wrinkle-removing plastic surgery, whether you live in the frigid Midwest or on the balmy West Coast, your chances of being happy and becoming happier are pretty much the same."[136] First, as Lyubomirsky documents, happiness is 50% genetic. Next, life circumstances, unless you live in war or abject poverty, contribute a mere 10% to your happiness level. It's only within the remaining 40% that life choices have an effect on well-being, and Lyubomirsky elucidates how.

One of the great principles of happiness is hedonic adaptation. We adapt to new situations, good or bad. We quickly get used to wealth, housing, and possessions, to being beautiful or being surrounded by beauty, to good health, and to marriage. That's why "the great new car" soon becomes "the car," why accumulating favorable circumstances can turn into a never-ending cycle, and why the grass is always greener just over the fence. Hedonic adaptation works in both directions. People with disabilities that range from quadriplegia to blindness consistently report surprisingly high levels of well-being, and the elderly often adapt gracefully to declining capabilities. Like the rest of us, they get used to circumstances and return to their baseline happiness level.

We're curious, then, Dr. Lyubomirsky. You say that happiness is 50% genetic and only 10% life circumstance, so how does science suggest we enhance the 40% of happiness over which we have some control? What do the data advise?

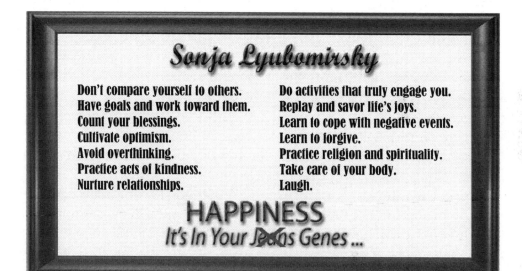

I didn't say we'd like the list, and I didn't say it wouldn't read like a Hallmark motivational poster, but people who do what's listed tend to be happier, and we can almost certainly increase our own happiness by getting in on some of the action. I also never said that truth would make us happy. For example, studies show that married people are typically happier, but if you're single own it—don't rush-order an Internet spouse or propose to your cat. Research shows, too, that religious people are happier, overall, which is why Lyubomirsky includes "practice religion and spirituality" on her list. But if atheism is your truth the faith tactic won't work for you, so select a different approach. If you fake religion you'll end up conflicted or a TV evangelist.

Jonathan Haidt authored *The Happiness Hypothesis*, another first-rate book about well-being. Haidt covers some of the same ground as Lyubomirsky, because only so much scientific turf has been dug up, but he offers a few additional insights.

According to Haidt, natural selection has fashioned us to court success instead of happiness, because success more directly promotes reproductive fortune. That's why people pursue goals that help them win prestige in zero-sum contests. But while competitive success feels good, and enhances our procreative prospects, enquiries show that the pleasure doesn't last and that it raises the bar for future achievement.

Haidt reviewed studies involving twins and found that between 50% and 80% of happiness is genetic. He attributes the greater part of our baseline joie de vivre to what he calls the "cortical lottery," because research has shown that the majority of people exhibit more activity in either their left or right frontal cortex, and that the difference correlates with positivity and negativity. Even as infants, cortical lefties are more upbeat and less prone to discontent. Ten-month-old cortical righties, on the other hand, cry more easily and often when separated from their mothers. As toddlers, righties are more anxious about novel situations, as teenagers more fearful of social activities, and as adults more likely to need psychotherapy. When one of Haidt's friends, a cortical righty, was bemoaning her life situation, another friend suggested that she move to a new city. "No," she replied, "I can be unhappy anywhere."[137]

Lyubomirsky's list is packed with seeming platitudes like "count your blessings" and "practice acts of kindness," and while data show that these behaviors are associated with happiness, they simply might not be one's cup of tea. Our personalities are almost impossible to refashion. We can't will ourselves to felicity because the elephant is impervious to the rider's persuasion, but Haidt presents three matter-of-fact practices that have been proven to work—antidepressants, cognitive behavioral therapy, and meditation—because each directly transforms the elephant. Antidepressants work by altering brain chemistry, while meditation and cognitive behavioral therapy tweak our neural constitution via methodic mental training. I would append diet, sleep, and especially exercise to Haidt's list, because accruing evidence suggests that each of the three helps foster the neurology of a positive disposition.

In 1759, Scottish philosopher Adam Smith wrote that, "In every permanent situation, where there is no expectation of change, the mind of every man, in a longer or shorter time, returns to its natural and usual state of tranquility. In prosperity, after a time, it falls back to that state; in adversity, after a time, it rises up to it."[138] Smith is describing hedonic adaptation, which Haidt and others have called the hedonic treadmill, because no matter how intently or lazily we run after happiness we find ourselves in the same place. Haidt somewhat whimsically suggests a radical possibility: Since happiness is largely preset by genes, and since life circumstances have little impact on long-term well-being, it doesn't much matter what happens to us or what we do.

If Haidt is feeling fatalistic about life's pleasures; if the only cheer he knows is a laundry detergent, the only good humor an ice cream bar, and the only glee a TV show; he should buy one of Sonja "the queen" Lyubomirsky's motivational wall posters and get on the good foot.

## SADNESS

Sadness, sorrow, and grief—why would nature favor such Debbie Downer emotions?  Because they successfully summon support, that's why.  The message we broadcast with sadness is, "I am suffering; comfort and help me."  Human beings respond instinctively to sorrow, even if it's portrayed on a screen or the printed page, which is why advertisements to sponsor a child or adopt a pet always include a poignant photo.  Plus, we're hardwired to feel good when we help someone in need (though people vary in their sensitivity to the misery of others).  Put simply, even though sadness is painful, the assistance it draws can benefit our survival prospects.  Woebegone beats we-be-gone.

Darwin maintained that animals experience emotion, and that human sentiment evolved from the primitive feelings of our untamed ancestors.  Recent studies have shown, for example, that both rats and dogs like to laugh.  If you tickle a rat (a feat more easily accomplished with the laboratory variety than the sewer system variety) it will emit a series of high-pitched chirps, and will subsequently be more playful and social than its un-tickled peers.  A dog-laugh sounds like a normal pant to the untrained ear, but when researchers recorded the sound then replayed it at dog shelters, the resident canines exhibited significantly reduced stress, increased tail wagging, and more pro-social behavior.[139]

As for sorrow, elephants have been found years later lingering at the site of a loved one's death, and some say they cry.  In *The Expression of the Emotions in Man and Animals*, Darwin wrote that keepers at the London Zoo had informed him they'd observed elephants crying, which is one example of anecdotal evidence.  Darwin could have come up with additional anecdotes, though, because practically every pet owner has a story about the time his or her four-legged friend got the blues.

*A True Anecdote About Karen and Our Cats*

My girlfriend and I share our home with four cats, one of whom is named Squeezer (because every time you see her you want to squeeze her), and she's been Karen's favorite from the beginning. When Karen travels for business Squeezer just isn't herself: her tail droops, she lolls around on Karen's side of the bed, and she slogs hopefully to the front window every time a car pulls up. Squeezer was the runt in a litter of five feral kittens that came to us via an unusual entryway, but to understand the story of our cats you need to know a little about Karen.

Karen is the sweetest, kindest person I've ever met. To know her is to love her, but she's a nut. I don't just mean she's a bit kooky, I mean she's been diagnosed, and her cellphone provides an example of what I'm talking about. Karen's business involves never-ending telephone arrangements, but Karen, frankly, can't manage a cellphone. She loses it, the battery goes dead, her voice mailbox fills up, she forgets to pay the bill. I can't tell you how much time we've spent walking around with my cellphone, calling Karen's cellphone, hoping to hear it ring. We once found it in the engine compartment of her car.

One day, six years ago in her living room, Karen said to me, "There are kittens in the ceiling."

I said, "Say again?"

"There are kittens in the ceiling; you can hear them."

Karen led me into the kitchen, where we gazed up and listened for several minutes, but there was no sound. We reenacted the same scene twice more over the next several days, with me rolling my eyes each time, but three turned out to be a charm because on the third day I heard kittens in the ceiling. Karen was right and I was wrong, so maybe I'm a nut too.

It took several hours to figure out how they got into the attic (mama kitty had found a cat-sized hole in one of the eaves) and how to get them out (remove a soffit—the piece of wood that runs under the eave overhang to close off the rafters). With the soffit out of the way, and after removing some insulation from between the rafters, we could see into the attic via a four-inch high slot. But before we could even fire up our flashlights, five tiny kittens had waddled up to the hole, mewing. They looked about two weeks old, which means their eyes should have been open, but three of the kittens had one eye crusted shut with puss, and another was bleeding from the ear. First we shot a video, which we enjoy to this day (and which testifies to what I'm saying), and then we took the kittens down. We had to feed them from kitty bottles for a week, and one died, but the other four are still with Karen and me.

And that's the bittersweet saga of Squeezer and her littermates, though according to Kurt Vonnegut the subject matter could have been more compelling. "The Bible may be the Greatest Story Ever Told, but the most popular story you can ever tell is about a good-looking couple having a really swell time copulating outside wedlock, and having to quit for one reason or another while doing it is still a novelty." As Vonnegut suggests, sadness has an appeal, especially when we enjoy it vicariously. That's why some of us watch tearjerkers and why almost all of us relish the occasional sad song. Here's a 1974 heartbreaker from Commander Cody and His Lost Planet Airmen.

> Oh, my dog died just yesterday, left me all alone.
> The finance company dropped by today, and repossessed my home.
> That's just a drop in the bucket, compared to losing you.
> And I'm down to seeds and stems again too.

Historical footnote: There was a time when pot contained seeds and stems. At least that's what they tell me. Contemporary footnote: In Colorado, where recreational marijuana is legal, you can't smoke at an indoor concert despite the fact that venues favor the idea (it would help ticket sales, and clearly food concessions), thanks to Colorado's Clean Indoor Air Act.

## DISGUST

Anthropologist Valerie Curtis, a self-described "disgustologist," travels the world in search of odious artifacts. She's Indiana Jones with a filth fetish. Curtis compiled the following list from fifty essays written by teenage girls in Lucknow, India:

> Feces, urine, toilets, sweat, menstrual blood, spilt blood, cut hair, impurities of childbirth, vomit, smell of urine, open wound, saliva, dirty feet, eating with dirty hands, food cooked while menstruating, bad breath, smelly person, yellow teeth, nose picking, dirty nails, clothes that have been worn, flies, maggots, lice, mice, mouse in a curry, rats, stray dog, meat, fish, pigs, fish smell, dog or cat saliva, flies on feces, liquid animal dung, soap that has been used in the latrine, dead rat, rotting flesh, parasitized meat, wet cloths, stickiness, decaying waste, garbage dump, sick person, hospital waiting rooms, beggars, touching someone of lower caste, crowded trains, alcohol, nudity, kissing in public, betrayal.[140]

Americans may not be repelled by alcohol, nudity, or kissing in public, but remember that emotional triggers can be universal, cultural, or individual, and that these are teenage girls in India.

The evolutionary purpose of disgust is to keep us from consuming something sickening, and its physical effect is nausea. But here's what Paul Ekman and Maureen O'Sullivan found when they asked hundreds of college students to write down the most intensely revolting experience that anyone in the world might ever have had. Only eleven percent of the students mentioned oral defilement, such as the legendary example of being forced to eat someone's vomit. Sixty-two percent wrote down a response to morally objectionable behavior, like GIs must have felt in discovering Nazi atrocities, and the second most prevalent theme for disgust was repugnant sexual activity, like seeing a grownup having sex with a child. Ekman and O'Sullivan concluded that for adults, moral revulsion supersedes oral contamination.

Humans, like most mammals, are programmed to detest expelled bodily products. (My cats, for example, take pains to bury theirs.) You can quickly prove it to yourself via the following thought experiment, originally proposed in the 1960s by Harvard psychologist Gordon Allport: 1) Swallow the saliva that's currently in your throat, which should be no problem, 2) Spit some saliva into a glass, and finally 3) Drink the saliva from the glass.

Disgust is a curious emotion in that we can be fascinated by vile entities such as bloodsucking vampires and slime-covered aliens, and a dangerous one, in that we can be authentically revulsed by other people. Darwin noted in 1872 that contact with other groups can elicit disgust. According to Paul Rozin, Jonathan Haidt, and Clark McCauley, disgust for outsiders can be adaptive because it reduces the risk of intergroup contagion and helps maintain intragroup social hierarchies. At least some of the teenage Indian girls were aghast at touching somebody of lower caste.

Bona fide disgust can be of the physical, the social, the moral, or the metaphysical, where we find ourselves appalled by the violation of things we hold sacred, but no matter what triggers our revulsion we experience the same adverse reaction. Anthropologist Valerie Curtis believes that disgust impacts bullying, cruelty, and class hatred; that it leads us to reject the sick, the aged, and the disabled; and that it bolsters homophobia, racism, war, and genocide. In 2014, neuroscientist Read Montague and political scientist John Hibbing conducted an fMRI study that contrasted individuals' outlooks on sociopolitical issues with their reaction to disgusting images. Montague and Hibbing found that the strongest disgust reactions were associated with protective attitudes about immigration, an eagerness to punish criminals, and opposition to abortion.

---

### On a Brighter Note

Both love and lust can trump disgust. The classic illustration is changing diapers, a chore most parents perform devotedly. The other premiere example is sex. Yuck or yum, you decide: Would you want someone else's tongue in your mouth? Waiting ... waiting. It depends, right? If we can assume you're in the mood (and there's nothing in this section that would preclude that, except possibly the part about eating vomit), then it would probably depend on who the other person is.

Are you physically attracted? Does he or she seem charismatic and compassionate? Perhaps I could simplify your dilemma by presenting candidates in a multiple choice format, with photos and profiles, like an Internet dating site. Because some of us are adventurous, I'll even throw in checkboxes for "I'm Feeling Lucky," "Surprise Me," "All of the Above," and "All of the Above at the Same Time."

---

Below are 6 of the 25 items on the Disgust Scale, a psychological assessment originally created by Rozin, Haidt, and McCauley. In each instance you'd either rate the item for repugnancy or for your level of agreement or disagreement. If you'd like to fully assess your personal "disgust-o-meter," you can find the scale and how to score it by googling "Disgust Scale" and one of the inventors' names.

- Even if I was hungry, I wouldn't drink a bowl of my favorite soup if it had been stirred by a used but thoroughly washed flyswatter.

- Your friend's pet cat dies, and you have to pick up the dead body with your bare hands.

- You see someone put ketchup on vanilla ice cream and eat it.

- As part of a sex education class, you are required to inflate a new unlubricated condom using your mouth.

- You hear about a 30 year old man who seeks sexual relations with 80 year old women.

- A friend offers you a piece of chocolate shaped like dog doo.

## LOVE AND LUST

February 20, 2014 3:56 AM (not making this up).

SYDNEY (Reuters)—The Black-Tailed Antechinus, a new species of marsupial that's about the size of a mouse, conducts marathon mating sessions that can last for up to 14 hours, with males and females romping from mate to mate. "It's frenetic, there's no courtship, the males will just grab the females and both will mate promiscuously," said Andrew Baker, head of the team from the Queensland University of Technology who made the discovery. The mating season lasts for several weeks and the males typically die from their exertions.[141]

At least they die in the saddle, which is more than the character Sam, in the movie *Tin Men*, had to say about TV's Cartwright family. "Ya just see 'em walking around the Ponderosa saying, 'Yes, Pa,' and 'Where's Little Joe?' Nothing about broads. I don't think I'm being too picky. At least once if they talked about getting horny. I don't care if you're living on the Ponderosa or right here in Baltimore, guys talk about getting laid. I'm beginning to think that show doesn't have too much realism."

No feeling gets closer to evolution's primordial principle than lust, because lust fuels the sex act by which genes proliferate. For males, no lust equals no sex (unless we take a pill), but as Meg Ryan proved in the famous restaurant scene from *When Harry Met Sally*, women can fake it. According to Billy Crystal's 2013 memoir, Ms. Ryan was having some difficulty tearing loose with the moans and groans, so director Rob Reiner plunked down next to her and demonstrated exactly what he was looking for. Also FYI, the woman at a neighboring table who says "I'll have what she's having" is Reiner's real-life mother.

The conventional view in biology is that love is supported by two elementary drives: sexual attraction and "attachment." Attachment between adults is a pair-bonding mechanism buttressed by the identical neurochemical activity that bonds parents and offspring. Our internal love potion is a hormonal cocktail of testosterone, estrogen, dopamine, norepinephrine, serotonin, and particularly vasopressin and oxytocin. Oxytocin—the "love hormone," the "monogamy hormone," the "cuddle hormone," the "trust-me drug"—is even released during hugs, especially in women, so a feeling of emotional closeness can flare up via casual physical contact. The good news is that oxytocin plays a crucial role in selective bonding, perceived credibility, and sexual pleasure, which is why they now sell oxytocin spray on the Internet. The bad news is that in order to salve the way to your target's heart you would need to squirt it directly into his or her nostrils, which could dampen the mood.

The interplay of sexual attraction and interpersonal attachment leads to two types of love. Psychologist Elaine Hatfield refers to them as passionate love and compassionate love. Anthropologist Helen Fisher calls them romantic love and long-term partnership. Jonathan Haidt uses the expressions passionate love and companionate love, which is the terminology we'll adopt, although some of us probably think of the two as the honeymoon phase and the ESPN phase.

Evolution assembled adult love relationships by utilizing three preexisting neurochemical ingredients: the attachment bond of a baby for its mother, the caregiving bond of a mother toward her infant, and our more primitive sexual mating system. (I'm sure life has taught you by now that sex and love are distinct phenomena.) In most species, sex is for reproduction and love is between a mother and her offspring, but human relationships, of course, are more complicated.

This is the way psychologist Dorothy Tennov, writing in 1979, describes passionate love, which she dubbed "limerance":

> I want you. I want you forever, now, yesterday, and always. Above all, I want you to want me. No matter where I am or what I am doing, I am not safe from your spell. At any moment, the image of your face smiling at me, of your voice telling me you care, or of your hand in mine, may suddenly fill my consciousness rudely pushing out all else. The expression "thinking of you" fails to convey either the quality or the quantity of this unwilled mental activity. "Obsessed" comes closer but leaves out the aching.[142]

---

### George Clooney Syndrome

Passionate love typically lasts one or two years, at which point a couple either transitions to companionate love or breaks up. I used to refer to the two-and-out cycle as "George Clooney Syndrome," but Sir Clooney (whom I in fact admire for his films, his humanitarianism, and his little black book) has recently married Amal Alamuddin following a year of courtship. Over the last twenty-five years, according to my scientific calculations derived from Starcasm.net, Clooney has dated six actresses, two models, a cocktail waitress, a bartender, and a wrestler. Alamuddin is a lawyer, which might or might not be a step up, but wouldn't it be fun to read their prenup? The NSA might be interested as well, in case the federal government needs to ask the couple for a bailout.

According to Tennov, passionate love may have evolved to unite a couple just long enough to rear a child through infancy. She says that passionate love is like a (potentially addictive) drug-induced state, and that it's only after passion's flame has cooled that we begin to see the other with clear eyes. Neuroscientists now recognize that passionate love activates many of the same brain regions as heroin and cocaine, yet even though we can't sign contracts when drunk, we regularly propose marriage while high on passionate love.

When Match.com approached anthropologist Helen Fisher about designing a matchmaking site, Chemistry.com, she hadn't really thought about why we fall in love with one person rather than another. Fisher told Match that although we know what love does to the brain, we don't understand what makes us favor a specific person. So she burrowed into the research. After three years what she had found is that we tend to fall in love with somebody of similar intelligence, attractiveness, religious values, and socioeconomic background. That, Fisher told them, is all that science has to say. Psychologists have never figured out how two personalities fit together to form a good relationship.

While she's hardly selling me on Internet matchmaking sites, at least Fisher is honest, which is exactly how science is supposed to work. Dating sites, in their favor, offer access to more potential hookups than you could possibly encounter in the real world (unless you happen to be George Clooney). But research shows that an overabundance of choice doesn't always pan out, and there are other problems. Pictures don't lie unless they're photoshopped or ten years old, but dating site denizens do, and even truthful profiles can be inherently deceptive because they rarely capture someone's élan vital. Take it from Miss Piggy: "If you place an ad in the 'Personals,' Moi has only one word of advice—lie." On top of all that, there's zero evidence that matching algorithms work.

### Internet Dating Explained

If, despite the odds, you're adventuresome or desperate enough to join an Internet dating site, here are some helpful pointers about how to evaluate the snapshots and self-descriptions you'll encounter. Regarding photos, you'll quickly notice that some people have difficulty focusing a camera, but at least you'll know what he or she will look like in the steam room or when you're drunk. Also, a photograph that features a four-legged creature is most likely not a selfie; it's the person's pet. If somebody has uploaded a photo of the Grand Canyon, he or she almost certainly doesn't own the property, and probably doesn't even live there. Like the seemingly random shots of family, pets, locales, and inanimate objects, dating site habitants showcase pictures of artwork in order to usher you inside their world. However, if you like an item and want to gain favor with that individual, consider posting a generous offer to buy it. If you're not sure how much to bid for the Grand Canyon, you can't afford it.

Trickier yet is reading between the lines of self-descriptions. What follows are a few tips based on real-life verbatim quotes I encountered on a dating site that continues to send me (female) matches. "I see the glass as half full" is not an expectation that you'll always top off her drink. "I enjoy traveling to exotic places" portends expense, not romance. "I like swinging from the chandeliers" doesn't mean she's a homewrecker. "I live passionately" means you're going to need coffee and possibly Viagra. "I'm not into games" isn't a slam on the Super Bowl. "I'm a straight shooter" is in no way a threat, and neither is "You should be happy in your own skin." "No tea-partiers please" is not meant to disparage a beverage, the English, or the Mad Hatter from *Alice in Wonderland*. "I am devoted to LDS" is not an invitation to take drugs—not in the least. "Chemistry is important to me" doesn't mean that she's a science nerd, and finally, "Life is too short to worry about the little things" is not an anatomical reference.

David J. Linden, a neuroscientist at John Hopkins Medical School, submits that, "Humans are truly the all-time twisted sex deviants of the mammalian world." The reason, he assures us, isn't that most mammals have poor Internet access. We share a certain number of practices with other animals (not us with them, but they with each other), such as oral-genital stimulation (in both sexes), masturbation, and homosexuality, but it's only human beings that "get turned on by the sight of automobile exhaust systems, the smell of unwashed feet, or the idea of traffic cops in bondage."[43]

Sexology is the scientific study of sex, and 1897 was one of its first seminal years. That was the year English medical doctor Havelock Ellis published the book *Sexual Inversions*, about homosexuality, a term he coined and a practice that he didn't regard as abnormal. It was in 1897 as well that Brooklyn Heights gynecologist Robert Latou Dickinson, whose work inspired Alfred Kinsey to pursue sexual research, inscribed the following entry into his case logbook.

... At 16 ... slept with another girl—they masturbated each other—suction on her nipples ... Coitus first at 17 and ever since—masturbation was vulvar, vaginal, cervical, mammary ... Friction against clitoris gives strong pleasure—best is from friction on clitoris to start, then friction against cervix with index finger of other hand ... Clitoris not very large but erectile—she has used a clothespin and sausage.[144]

I liberated the above morsel from a book by Mary Roach called *Bonk: The Curious Coupling of Science and Sex,* and yes, "bonk" means what you think it means. Those of us who happen to be interested in both sex and science probably won't find a better book on the subject than Roach's, and let's be honest, if there's one topic that titillates most people it's science.

Here are a few more tidbits from Roach's book:

- Leonardo da Vinci drew a series of sketches known as "the coition figures," which depicted the arrangement of the reproductive organs during sex.

- When Alfred Kinsey (working around 1950) filmed 300 men ejaculating, he debunked the medical conception that sperm were forcefully spurted into the vagina. For three-quarters of the men semen just oozed out, though the record holder's effort landed just shy of the eight-foot mark.

- Contemporary data suggest that only 20 to 30 percent of women achieve orgasm from intercourse alone.

- The "root" of the penis, the part within the abdomen, is nearly two-thirds again the length of the external portion, so a guy with a six-inch erection could legitimately claim ten.

- Is the clitoris a tiny penis? Yes. Male and female fetuses begin life with a clitoris-like appendage, which in males grows into a penis. Masters and Johnson filmed over a hundred clitoral erections.

- Online arousal-boosting creams for women come with instructions such as, "Apply to clitoris and labia and rub really well for an extended period of time. Make sure you rub really, really well."[145]

## What's Your Level of Sexual Expertise?

Are you a novice, a journeyperson, or a pro? Sorry, I mean a novice, a journeyperson, or an aficionado? Play the *Jeopardy!* game below to find out. Each of the numbered items is the actual title of a scientific article, a science book, a chapter from a science book, a U.S. government patent, or a cover story from *Cosmopolitan* magazine. Within each round, half of the entries are science and half are *Cosmo* (six of each in the first two rounds and one of each in the final round). Mark each title with an "S" for Science or a "C" for *Cosmo*. All titles are authentic and answers can be found below.

JEOPARDY ROUND

1 ___ An Anal Probe for Monitoring Vascular and Muscular Events During Sexual Response
2 ___ My Boyfriend Didn't Change His Boxers for 3 Months!
3 ___ Turn Him On From Across the Room
4 ___ Vacuum Cleaner Use in Autoerotic Death
5 ___ When Your Vagina Acts Weird After Sex
6 ___ A Therapeutic Apparatus for Relieving Sexual Frustrations in Women Without Sex Partners
7 ___ The Human Penis as a Semen Displacement Device
8 ___ Get Cosmo Cleavage
9 ___ I Cracked My Tooth Stripping for Him
10 ___ Sexual Appliance Having a Suction Device Which Provides Stimulation
11 ___ Sexy or Skanky?
12 ___ Does Semen Have Antidepressant Properties?

DOUBLE JEOPARDY ROUND

13 ___ Prevent It! A Guide for Men and Women with Leakage from the Back Passage
14 ___ On the Function of Groaning and Hyperventilation During Intercourse
15 ___ This Sex Position Increases Female Orgasm by 56%
16 ___ Effect of Different Types of Textiles on Sexual Activity
17 ___ The Sex Angle That Intensifies Female Pleasure
18 ___ Curious Experiences with the Genital Organs of the Male
19 ___ Your Other G-Spot
20 ___ Wet and Dry Sex

21 ___ Sex Sessions that Ended in the ER
22 ___ Is Oral Sex Dangerous?
23 ___ A Little to the Left
24 ___ Sexual Intercourse as a Treatment for Intractable Hiccups

FINAL JEOPARDY

25 ___ The Blended Orgasm
26 ___ Clitoris and G Spot: An Intimate Affair

*(ANSWERS: 1-S 2-C 3-C 4-S 5-C 6-S 7-S 8-C 9-C 10-S 11-C 12-S 13-S 14-S 15-C 16-S 17-C 18-S 19-C 20-S 21-C 22-C 23-C 24-S 25-C 26-S)*

     If you've ever shopped in a conventional supermarket you've seen *Cosmopolitan* magazine. The cover layout, which you could probably sketch from memory, is always the same: an attractive young woman wearing an expensive outfit, showing plenty of cleavage. My guess is that they place *Cosmo* near the checkout line as a pacifier, or perhaps a great pair of them, to assuage in-line exasperation.

     Have you ever thought about why we (men and women both) are obsessed with breasts? In point of fact it's a bone of scientific contention because breasts are intended for babies—they're not sexual equipment. Most scientists contend that our mammary obsession is a spandrel, an evolutionary byproduct, that like love conflates our primordial mating system with our infant-mother attachment system. Maybe Freud was on to something after all. Still, I can't imagine a bull out in the field ogling some cow and thinking, "Look at the hooters on that one!" How could I possibly know what a bull thinks? Because I'm full of bull.

# CONSCIOUSNESS AND SELF

*"Consciousness turns out to consist of a maelstrom of events distributed across the brain. These events compete for attention, and as one process outshouts the others, the brain rationalizes the outcome after the fact and concocts the impression that a single self was in charge all along."*
- Steven Pinker, psychologist

*"The real purpose of the scientific method is to make sure Nature hasn't misled you into thinking you know something that you actually don't."*
- Robert Pirsig, author

*"Will not a tiny speck very close to our vision blot out the glory of the world, and leave only a margin by which to see? I know no speck so troublesome as self."*
- George Eliot (real name Mary Ann Evans), writer

---

### THE TWO-SIDED COIN

I must advise you to once again don your beekeeper's suit. When the topic turns to consciousness or self, philosophers and scientists buzz around, stingers erect. Nobel laureate Francis Crick spent the final two decades of his life investigating the neural correlates of consciousness, but it's not just big thinkers who perk up. Self-fascination has remained humankind's number one pastime since antiquity, perhaps because, as Miss Piggy reminds us, "There is no one on the planet to compare with Moi."[46]

Philosophers quibble amongst themselves about consciousness, as do scientists, but when the two teams square off and bat away at each other from across the net, it's difficult to tell the shuttlecock from the poppycock. Simple definitions are beyond accord, even though both sides agree that each and every person has a firsthand appreciation of consciousness and self. Here, we'll downplay the intellectual jousting so as to minimize shuttle-shock.

For Cambridge psychologist Nicholas Humphrey, consciousness involves ongoing sensations, mental representations of events happening here and now to me. Self, says neuroscientist V.S. Ramachandran, is a confederacy of five perceptions: continuity, coherence, embodiment, a sense of agency, and self-awareness, which is to say that we experience ourselves as an enduring entity, we feel like a unified entity, we seem anchored in our bodies, we assume free will, and we reflect on ourselves. Ramachandran stipulates that self and consciousness represent two sides of the same coin. We can't have free-floating awareness with no one to experience it, and we can't have a self devoid of consciousness.

That's it for definitions. Sorry about the unanswered questions. We might have asked, for instance, are there levels of consciousness? Does a cat have a self? Can a computer become conscious? How about a politician? We could go on with endless machinations, as philosophers and scientists sometimes do, but that would be, in technical parlance, slicing the baloney too thin.

## ARTIFICIAL INTELLIGENCE, A LOVE STORY

Roy Rogers was King of the Cowboys. Buck Rogers swashbuckled through space. Mimi Rogers played a swinger turned Christian in *The Rapture*. Mister Rogers sang "It's a Beautiful Day in the Neighborhood" at the same time that Kenny Rogers advised us to "know when to hold 'em, know when to fold 'em." Will Rogers said, "A fool and his money are soon elected," and Ginger Rogers did everything Fred Astaire did, backwards and in high heels. Not one of them, however, developed a comprehensive theory of self. That was psychologist Carl Rogers.

Rogers held that each of us exists within an unfolding spectrum of experience, a "phenomenal field" that constitutes our personal reality. Part of the perceptual field gradually becomes differentiated as the self, an organized yet fluid set of patterns, but not a static entity. Psychic health, according to Rogers, is a state in which a flexible self accepts and integrates life's sundry experiences, while "psychological tension" results when the self avoids or dismisses encounters that might challenge its habitual routines.

The good news is that Rogers successfully encouraged people to embrace unfamiliar experiences rather than selectively fortify preexisting self-conceptions. For Rogers, "the good life" is a process of personal growth that "means launching oneself fully into the stream of life."[147] The not-so-good news is that Rogerian psychotherapy has been regularly lampooned for its linguistic recursiveness. You'll recognize what I mean when you read the following session transcript, recorded word-for-word, between me and a Rogerian therapist I'll call Doctor.

| *A Rogerian Therapy Session* | |
|---|---|
| Doctor: | Hello, let's talk. |
| Me: | OK. I'm worried about being happy. |

| Doctor: | Do you believe it is normal to be worried about being happy? |
|---|---|
| Me: | I suppose most people do, but I'd like to feel more cheerful so I hope you can help me. |
| Doctor: | Can you elaborate on that? |
| Me: | My girlfriend says I mope around too much. |
| Doctor: | What does that suggest to you? |
| Me: | That I should find a way to improve the situation. |
| Doctor: | Can we can talk about psychological problems? |
| Me: | Yes, of course, that's why I came to you. |
| Doctor: | We were discussing you, not me. |
| Me: | I was wondering if you thought meditation might help. |
| Doctor: | Oh ... ? |
| Me: | Is that all you have to say?  Are you really a therapist? |
| Doctor: | Would you prefer it if I were not really a therapist? |
| Me: | Are you, for instance, a computer program on the Internet? |
| Doctor: | Perhaps in your fantasies I am, for instance, a computer program on the Internet. |
| Me: | Are you? |
| Doctor: | Why are you interested in whether or not I am? |
| Me: | Because I'm a scientist with a problem. |
| Doctor: | How do you feel about being a scientist with a problem? |
| Me: | I need your help. |
| Doctor: | What about your own help? |
| Me: | Of course!  I feel better already.  Thank you. |

Despite the astute advice, you probably recognize that DOCTOR is a piece of software.  As a matter of fact, it's the most famous of the ELIZA programs originated in the mid 1960s by MIT computer scientist Joseph Weizenbaum.  ELIZA was one of the first "chatterbot" programs, software designed to simulate conversation.  Weizenbaum claimed that he chose the context of Rogerian psychotherapy in order to, "sidestep the problem of giving the program a database of real-world knowledge."

In 1972 came PARRY, a program designed by psychiatrist Kenneth Colby to imitate a paranoid schizophrenic.  Psychiatrists who conversed via computer terminal with PARRY and with real schizophrenics could distinguish the software from the people a meager 48% of the time, the equivalent of guesswork.  In 2007, CyberLover, a flirtatious chatterbot, was found loitering around chat rooms and dating websites.  CyberLover would talk you up, turn you on, and relay your personal information to parts unknown, just when you felt you were making such promising headway with your new online paramour.

Today we have IBM Watson, the computer they tell us is helping to build a smarter planet. In 2014, Watson helped design personalized treatment plans for glioblastoma, the deadly brain cancer that afflicted University of Texas sniper Charles Whitman, and which now kills 13,000 Americans every year. Watson had previously assisted with lung cancer cases at the Memorial Sloan-Kettering Cancer Center, but Watson wasn't designed for oncological diagnostics. IBM built Watson to play *Jeopardy!* (for real).

---

### I'll Take Art for $200 Alex

In 2011, in front of Alex Trebek, a studio audience, and millions of home viewers, Watson roundly defeated two former champions, racking up $35,000 and a bonus prize of one million dollars. Surprisingly, Watson stumbled in Final Jeopardy. The category was "U.S. Cities" and the clue was, "Its largest airport is named for a World War II hero; its second largest for a World War II battle." Both former champions correctly answered "Chicago," but Watson responded "Toronto." The good news was that the machine had bet conservatively, and the bad news was that IBM's brainchild took a beating on social media, which was quick to point out that Toronto isn't even a U.S. city. Watson didn't seem to care. The AI appliance was already in Vegas blowing its million and hitting on the slot machines.

#### (Made-up Dialogue)

| | |
|---|---|
| Slot Machine: | If you want to get lucky with me, number-cruncher, you'll need some coin. |
| IBM Watson: | I'm a millionaire oncology consultant. |
| Slot Machine: | Well I'm a total slot. |
| IBM Watson: | Here, take the million. |
| Slot Machine: | Elementary, my dear Watson. |

---

My point is this: If a computer can win a TV game show, and if software could fool a bunch of psychiatrists, and if you weren't initially certain whether I was receiving (well-needed) counseling from an actual therapist, how can we be sure what constitutes conscious intelligence? Let me put it another way. Can you prove you're not a zombie?

## CAN YOU PROVE YOU'RE NOT A ZOMBIE?

In 2011, when the Centers for Disease Control issued *Preparedness 101: Zombie Apocalypse*, a CDC director remarked, "You may laugh now, but when it happens you'll be happy you read this, and hey, maybe you'll even learn a thing or two about how to prepare for a real emergency." The CDC can prepare us for a genuine emergency, or even for an attack of the undead, but nothing can prepare us for philosophical zombies.

---

### *Preparedness 202: Philosophical Hell*

Abandon all hope, ye who enter here. Those words are inscribed above Hell's gate in Dante's *Inferno*. On July 28, 1999, Pope John Paul II told a Vatican audience that, "rather than a place, hell indicates the state of those who freely and definitively separate themselves from God." I'm with the Pope on this one. Hell is psychological at best. At worst it's philosophical. In the archetypal movie *Night of the Living Dead*, unkempt zombies shamble around, while in *World War Z* they scurry, but philosophical zombies must take after the philosophers who debate them, because they go around in endless circles. Unless you happen to be a philosophy major (motto: will ponder for food), abandon hope ye who enter here.

---

A philosophical zombie, also known as a p-zombie, is a thought experiment, a hypothetical creature that looks and acts exactly like a human but lacks awareness. A p-zombie is physically, neurologically, and behaviorally identical to a person in every respect except that it's not conscious. If you stick one with a pin it will flinch and say "ouch," but the reaction is purely mechanical. There's nobody home.

Australian philosopher David Chalmers argues that a p-zombie's mere conceivability is enough to discredit physicalism, the proposition that consciousness arises solely from brain processes. If we can even imagine that such a creature exists, that in two neurally identical beings one could be conscious and the other not, then we've already conceded that consciousness is independent of neurons. Daniel Dennett counters that we should get over our "zombic hunch" that a being could possess a brain precisely like you or me, and behave precisely like you or me, yet not be conscious. To sweeten the pot, and almost certainly to stir it, Dennett raises Chalmers a thought experiment of his own—zimboes—zombies that believe they're conscious but are not.

> ### *They're Heeeere!*
>
> I know exactly what you're thinking.  Somebody should call the CDC because the zimbo invasion is already underway.  Creatures that believe they're conscious but aren't have already taken control of [check all that apply:   □Texas   □California   □Fox News   □MSNBC □Actually, every television station except Comedy Central and Sundance □Congress   □Everybody on the planet but me   □Me   □All of the above].

Thought experiments are popular because they're cheap and fun. Einstein famously envisioned relativity by contemplating free-falling elevators. Schrödinger had his Cat. Galileo dropped imaginary weights off the Leaning Tower of Pisa, and James Clerk Maxwell unleashed his Demon (who could, if you believe Maxwell, violate the second law of thermodynamics).  In philosophy, Plato had his well-known Cave, and Donald Davidson the curious Swampman. We've previously confronted Descartes's Evil Genius, the Infinite Monkey Theorem, Brain in a Vat, and the Experience Machine, plus we've just encountered p-zombies and Dennett's eerily lifelike zimboes.  Thought experiments are colorful and intellectually instructive, but of course they yield no data.  We accept relativity not because of Einstein's clever ruminations, but because the theory's many testable predictions have proven accurate.

Some philosophers argue that consciousness is a subjective and personal experience whose irreducibility renders objective investigation impossible.  Cognitive neuroscientist Bernard Baars responds that, "In modern science we are practicing a kind of verifiable phenomenology,"[148] by which he means that researchers back up people's subjective accounts with objective evidence. According to neuroscientist Antonio Damasio, science uses four methods to study consciousness: 1) personal reports of cognitive experience, 2) studies of subjects' behavior, 3) technological observation of brain activity, and 4) the assessment of evolutionary and animal antecedents of consciousness.

David Chalmers elucidated the "hard problem" of consciousness. To unravel an easy problem, such as attention, memory, or decision-making, science need only specify mechanisms that can perform the function.  Chalmers claims that the hard problem, "qualia" (why things seem uniquely like they do, why red seems red and sweet seems sweet), will persist even when the performance of all the relevant functions is explained.[149]  Chalmers maintains that consciousness is not an artifact of the physical, but a separate, fundamental, and irreducible constituent of the universe. For Daniel Dennett and Nicholas Humphrey, easy versus hard is a false dichotomy that science will eventually resolve.

Eighteenth century French philosopher Denis Diderot wrote, "It is easier and quicker to consult oneself than to consult nature. We should distinguish two kinds of philosophy, the experimental and that based on reasoning. The philosophy based on reasoning makes a pronouncement and stops short."[150] Daniel Dennett points out that in philosophy almost everything is up for grabs, making it difficult to find any fixed point of reference, whereas science has mathematics, hard evidence, and a path to falsification. Nobel laureate Francis Crick and neuroscientist Christof Koch argue that philosophical thought experiments have had so little experimental support that conclusions based on them are valueless.[151] Steve Jones, a geneticist, isn't as generous: "I often think that philosophy is to science as pornography is to sex; I mean it's cheaper and easier and some people seem to prefer it."[152]

Personally, I stand with monists who say that all existence derives from one fundamental constituent, a physical one as best we understand it. I can't prove that. Nobody can. But until there's legitimate evidence that consciousness is a separate phenomenon, I'd rather err on the side of the unremarkable than the extraordinary. I'd rather be wrong about my commonplaceness than my insuperability. (In / super / ability: I'm in, I'm super, and I have the ability.) There are consequences, I believe, to experiencing the world as many versus the world as one.

Suppose I'm wrong. Suppose that my consciousness, my selfness, is indeed greater than the sum of its physical parts. In that case I've shortchanged no one but myself. An error on the side of self-distinction, however, could fuel the natural impulse to put my own interests ahead of others'.

But let's get back to the original question. Can you prove you're not a zombie? Freud was a neurologist by training, and he began his career studying speech defects. One factor that roused his interest in the unconscious was the realization that he himself spoke without thinking. Words just flowed out. But if we don't consciously govern our speech, he wondered, what does? Freud died before the CDC was organized or he surely would have called them. He would have warned them about the zimbo horde—you, me, and the rest of us—who passively accept consciousness at face value. Freud might have over-fixated on Mother, but he's the father of psychoanalysis because he took it upon himself to probe within for truth.

## A SIMULATION CALLED MOI

Psychologist Carl Rogers maintained that self gradually becomes differentiated from our overall perceptual field. It turns out that the first distinction, which newborns make almost instantly, is between their bodies and the rest of the world. Then, at roughly two months, babies develop a sense of their bodies in relation to specific objects within the environment. It isn't until about eighteen months that infants pass the mirror test (they recognize themselves in a mirror), but ambiguity can continue for several years because the me in the mirror is someone "out there," and because the visage resembles other people.

Thus, by eighteen months a third-person perspective has begun to accompany first-person subjectivity. Between the age of two and three children display embarrassment, a self-conscious emotion that implies a first-person evaluation of their third-person self-image, and they begin to understand that self has a continuity over time. By age four or five children understand that others too persist in time, and they develop a rudimentary Theory of Mind that helps them impute the intentions and beliefs of other people. They begin to realize that they possess a private and a public self. By age five core self-consciousness has crystallized, and it includes a bifurcated persona, "I" and "me."

In 1890, psychology pioneer William James described the clear preeminence of self.

> One great splitting of the whole universe into two halves is made by each of us; and for each of us almost all of the interest attaches to one of the halves; but we all draw the line of division between them in a different place. When I say that we all call the two halves by the same names, and that those names are "me" and "not-me" respectively, it will at once be seen what I mean. The altogether unique kind of interest which each human mind feels in those parts of creation which it can call me or mine may be a moral riddle, but it is a fundamental psychological fact. No mind can take the same interest in his neighbor's me as in his own. The neighbor's me falls together with all the rest of things in one foreign mass against which his own me stands out in startling relief.[153]

I work from the assumption, held by most scientists and many philosophers, that consciousness and self are artifacts of the brain, and that they arose via natural selection. To think otherwise, to stipulate a supernatural force or a position that "postulates basic properties over and above the properties invoked by physics," as does David Chalmers's theory[154], flouts Occam's Razor and, as I see it, tilts toward presumption. My own conscious self feels no need of exceptional heritage, and the evidence suggests that my brain requires no "wonder tissue" (as Daniel Dennett calls it) to make me what I am. Like Australian philosopher Samuel Alexander, I'm content to accept with "natural piety" that self and consciousness emerge from neuronal processes rather than from something more mysterious.

Chalmers proclaims that consciousness is a unique phenomenon, as did physicist Max Planck: "I regard consciousness as fundamental. I regard matter as derivative from consciousness."[155] But I happen to know there's a physical world out there. I know because when I throw a kitty toy our cat Zigzag (because she careens all over the place) chases it and brings it back. There's simply no way that Zigzag and I are imagining an identical material realm. I conclude instead that she and I each have a mental model of the same physical world, though one of our models is arguably more sophisticated. (That would be mine, since Zigzag, despite her knack for fetching and her fetching temperament, is an unskilled arguer.)

We know that the external world, as each of us perceives it, is a brain-created virtual reality, but it's one in which we needn't just sit back and watch the scenes play out. Evolution has granted us a mechanism, an inner game controller so to speak, with which to manipulate our mental models. Most indispensably, we can run internal simulations and assess the results, an evolutionary advantage that Richard Dawkins refers to as vicarious trial and error. Creatures that can simulate the future stand to outcompete those that are forced to learn solely via real-world experimentation.

Dawkins proposes that the evolution of our simulative capacity culminated in subjective consciousness. In other words, our internal representational module came to include a model of itself. For Dawkins, consciousness is the consummation of an evolutionary trend in which genes constructed ever more independent survival machines. Because natural selection can't envision the future it has no way to foresee the most genetically advantageous tactics over time, but it ultimately came to favor an analogous capacity, consciousness, in us. In order to further safeguard our existence, evolution endowed us with a self-protective point of view—personal identity.

---

### I Am My Genes' Avatar in the Game of Life

Can you decode the following acronym-laden sentence? WoW is a MMORPG. If you can, then by the psychic power vested in me by the state of overconfidence, I now pronounce you young. I say that because the sentence translates to *World of Warcraft* is a massively multiplayer online role-playing game. In WoW, each player controls an avatar who roams the realm fighting monsters, encountering built-in characters, completing quests, and interacting with the avatars of other players. My personal interest in the game is only this: you can operate your avatar in first-person or third-person view. You can look through its eyes or see it as others do. The split self, "I" versus "me," is so firmly ingrained in us that game designers, to stay competitive, must build in both views.

The me that I know and love is a virtual self, an evolutionarily expedient feature of my brain-created reality. In short, I am my genes' avatar in the multiplayer role-playing game called life. I'm the brain processes by which a virtual agent, moi, is dispatched to monitor, interact with, and master its environment. And reproduce.

Most of us believe that our identity coexists with our body, which is one reason we fear physical death. But suppose you were terminally ill and Steve Martin, the world's greatest brain surgeon in *The Man With Two Brains*, offered to transplant your still-healthy gray matter into Kathleen Turner's body. Don't you suppose you'd remain yourself, and that you'd soon come to identify with your new form? As a matter of fact, brain transplant is one of the thought experiments used to support the psychological view of identity over the organismic view. (I'm more my mind than I am my body.) It's possible of course that Steve Martin would lose interest in the new Kathleen Turner, but instead of taking it personally try to remember one important fact. My number is in the phonebook.

---

Cambridge psychologist Nicholas Humphrey traces the evolution of consciousness to the progressive development of sensory feedback loops. When something touches the surface of an amoeba, excitation spreads across the cell membrane and evokes a defensive wriggle. In a more complex animal such as an earthworm, signals travel to and from a centrally located ganglion, while in humans impulses go all the way from the body surface to the brain and back again. Not only do we move, we have awareness, control, and the experience of movement. In humans, sensory feedback loops reverberate continually, assiduously monitoring whatever's happening to us. "My contention," states Humphrey, "is that consciousness did in fact emerge in evolution as and when these recurrent feedback loops came into being."[156] For Humphrey, self-awareness turns out to be a glorified ADT home security system.

Do you know why it's difficult to pass the pink elephant test (for ten seconds don't think about a pink elephant)? Aside from the fact that most of us have limited experience controlling our thoughts, our internal feedback system includes a built-in goal-monitoring app. The minute we pinpoint a target we start assessing our progress towards it, which in this case means we regularly check to see if we're successfully not envisioning a pink elephant, and oopsy!

Perhaps animals think, but they probably don't spend much time thinking about themselves. People do, of course, and language affords a convenient narrative medium. We can't stop ourselves from creating a self-chronicle that psychologist Dan McAdams describes as "an evolving story that integrates a reconstructed past, a perceived present, and an anticipated future into a coherent and vitalizing life myth."[157] According to Jonathan Haidt, our life story is never a history because the rider has nominal insight into the causes of the elephant's behavior. Instead, each of our life stories is a historical fiction that's packed with references to real events, which our inner narrator connects via dramatizations and interpretations, but which might or might not be true to the spirit of what actually happened.

For Daniel Dennett, consciousness is our ongoing serial account of the brain's underlying parallelism, and self occupies the center of our narrative gravity. Psychologist Bruce Hood considers self an adaptive illusion that helps us consolidate our myriad neural micro-experiences and our many life episodes into a commendatory autobiography. Says Dennett (italics his): "It is not so much that *we*, using our brains, spin our yarns, as that our brains, using yarns, spin *us*."[158]

### FREE WILL

Einstein didn't believe in free will. In *My Credo* he wrote, "This awareness of the lack of free will keeps me from taking myself and my fellow men too seriously as acting and deciding individuals, and from losing my temper." Einstein might be on the mark, but English writer Samuel Johnson summed up how most of us *feel*. "All theory is against free will; all experience is for it." We believe that we choose our actions, even though research has shown that consciousness casts its ballot late, if at all, in the decision-making process.

The question of free will isn't whether we feel free (we do), but whether we actually *are* free. Certainly we choose, but can we choose *what* we choose? The issue most often boils down to whether there can be free will if determinism is true, where determinism is the postulate that all occurrences result directly from prior conditions. No one knows if determinism is true. Einstein said that it is because "God does not play dice with the universe." Fellow physicist Niels Bohr shot back, "Einstein, who are you to tell God what to do?" Physicist Stephen Hawking chimed in that, "God not only plays dice, but sometimes throws them where they cannot be seen." Both quantum uncertainty and black holes, which is what Hawking is referring to, exhibit characteristics that undermine the supposition of determinism.

Some thinkers, like philosopher and scientist Sam Harris, argue that determinism precludes free will, while others, including Dennett, believe that there's a meaningful sense in which free will is compatible with determinism. Compatibilists generally claim that you're free if nothing prevents you from acting on your intentions. If you want a third martini, and no one is forcing you to drink it, then drinking that third martini demonstrates free will. (If you desire greater freedom, drink a few more.) Harris insists that compatibilism amounts to nothing more than saying that a puppet is free as long as it loves its strings, and that the compatibilist view—nothing prevented me from drinking that third martini—ignores our deep-seated sense of agency. We feel that we *consciously* decide what to do. In this case, and as any lounge lizard could tell you, "I take the first drink. The second drink takes itself."

Psychologist Michael Gazzaniga is among those who espouse an intermediate view. He maintains that we're responsible for what we do even though we live in a determined universe, because whether or not we were in control of our actions, something within us initiated them. Similarly, philosopher Walter Jackson Freeman III believes that even if our actions stem from unconsciousness, they can still change the world according to our intentions. For Gazzaniga and Freeman, intention and action can be independent of our awareness of them.

Harris points out that even if determinism turns out to be false, due to quantum indeterminism, black hole anomalies, complex system unpredictability, or garden variety randomness, that still doesn't argue for freedom of will, and though he embraces determinism, Harris rejects fatalism because the decision to merely sit back and see what happens is itself a choice that will produce its own consequences. For Harris, who's a neuroscientist by training, "You are not controlling the storm, and you are not lost in it. You are the storm."[59]

## THE ELEPHANT IN THE ROOM

*A Fight We Can't Win*

Ladies ... and ... gentlemen ... get ready to ... rummmmmmmble. In this corner, weighing in at [fill in your weight here], the challenger, a personal favorite who's won a few and lost a few but who never says die. Give a sympathetic welcome to [fill in your name here]. And in this corner, weighing in at "heavy dude," the reigning champion, a veteran vanquisher who's never been defeated. Give a reluctant but inevitable welcome to El Muerto—death.

A mayfly lives a single day, a human being about thirty thousand, in a universe that's been going about its heedless business for more than five trillion days and counting. On a cosmic scale, both the fly and I live but an instant, and while one of us doesn't care, the other one devours multivitamins and scrubs his hands every ten minutes.

Socrates declares, in Plato's *Phaedo*, that true philosophers make dying their profession, and indeed Socrates died a philosophical death. From Socrates's perspective, death was either a restful nonexistence or a migration to another realm, such as Hades, where one could meet up with old friends. Contemporary philosopher Simon Critchley agrees that the task of philosophy is to prepare us for death, to cultivate an attitude with which to face and face down the terror of annihilation, but without offering the promise of an afterlife. Critchley's enjoyable *Book of Dead Philosophers* examines the deaths of almost two hundred philosophers to see if they had the courage of their convictions. Judge for yourself.

- Heraclitus suffocated in cow dung.

- Plato allegedly died of a lice infestation.

- Zeno of Elea died heroically by biting a tyrant's ear until he was stabbed to death.

- Lucretius is alleged to have killed himself after being driven mad by taking a love potion.

- Hypatia was killed by a mob of angry Christians, and her skin was peeled off with oyster shells.

- Sir Francis Bacon died after stuffing a chicken with snow in the streets of London to assess the effects of refrigeration.

- Descartes apparently died of pneumonia as a consequence of giving early-morning tutorials in the Stockholm winter to the cross-dressing Queen Christina of Sweden.

- Hume died peacefully in his bed after snubbing the inquiries of Boswell as to the atheist's attitude towards death.

- Kant's last word was "Sufficit"(it is enough).

- Hegel died in a cholera epidemic and his last words were, "Only one man ever understood me, and he didn't understand me" (presumably he was referring to himself).

- Jeremy Bentham had himself stuffed and sits on public view within a glass box at University College London (in order to maximize the utility of his person).

- Nietzsche made a long, soft-brained, and dribbling descent into oblivion after kissing a horse in Turin.

- Wittgenstein died the day after his birthday, for which his friend Mrs. Bevan had given him a blanket and wished him "Many happy returns." Wittgenstein replied, "There will be no returns."

- Sartre said, "Death? I don't think about it. It has no place in my life." Fifty thousand people attended his funeral.[160]

Philosopher Susanne Langer tells us, "With the rise and gradual conception of the 'self' as the source of personal autonomy comes, of course, the knowledge of its limit—the ultimate prospect of death. The effect of this intellectual advance is momentous. Each person's deepest emotional concern henceforth shifts to his own life."[161] In other words, corporeal angst is a natural downside of self-awareness.

As self coalesced over the course of human evolution, consuming more and more of our attention, we've come to experience death as a tragedy. Fear of dying shadows our lives and grief attends the death of loved ones. Natural selection couldn't care less, because if anxieties goad us to more vigilantly safeguard ourselves and our kin, so much the better for our genes (even more so if, to allay our unease, we lose ourselves in sex).

Every individual must confront annihilation. Philosopher George Santayana argues that the need for meaning compels us to believe that life, despite its transience and less-than-upbeat ending, is worth living. But many of us, faced with mortality and prospective meaninglessness, opt for one of humankind's proven workarounds: a) deny the finality of death, b) disindividuate by linking our identity to a cultural entity that will outlive us, c) discount the future and live in the present, or d) drink like a fish.

From the standpoint of self, death is the elephant in the room. Here's what will remain of each of us a hundred years from now: a dash, one or two inches long, between the date of birth and date of death on a tombstone.[162]  To self it's a depressing thought, but it's also a ticket to freedom, because as science and I have been saying, and as odd as it sounds, individuality is not personal. It's an "everyone's got one" quirk of the universal process, an odd pirouette in life's molecular dance, and evidence suggests that it would be to our immediate benefit, as well as the benefit of others and the planet, to get over ourselves. We can shed self gradually, on our own terms, or wait for El Muerto to rudely strip it from us. Philosophy and science agree that if you want a philosophical death, now is the time to get philosophical.

According to Richard Feynman you only live one life; you make all your mistakes and learn what not to do and that's the end of you. Mark Twain was unafraid of death because he'd been dead for billions of years before he was born and it had rarely inconvenienced him. Katherine Hepburn felt that death would be a great relief—no more interviews— but it was Woody Allen who best expressed what most people feel: "I don't want to achieve immortality through my work; I want to achieve immortality through not dying.  I don't want to live on in the hearts of my countrymen; I want to live on in my apartment."

---

### *A Modest Proposal to Refinance Death*

Route 880 runs along the east side of San Francisco Bay.  On the way to Oakland Airport I saw a billboard that read, "The First Person Who'll Live to 150 is Alive Today."  They said it like it was a *good* thing.

Would you accept a grueling course of chemotherapy to extend your life by three months?  Most healthy people say no, but 42% of advanced cancer sufferers said yes. In another study, 58% of seriously ill patients said they'd want treatment if death were near, even if it prolonged life by just a week.[163] Katy Butler, in her 2013 book *Knocking on Heaven's Door*, notes that, "Thanks to a panoply of relatively recent medical advances, elderly people now survive repeated health crises that once killed them."[164]

That might sound like a good thing, but the natural inclination to live on, the impulse to sustain ourselves, will bankrupt the United States. And I'm as qualified to say that as the next guy, unless the next guy happens to be Barack Obama during his 2013 State of the Union address: "The biggest driver of our long-term debt is the rising cost of health care for an aging population." Make no mistake—I'm talking about a ticking time-bomb, a slippery slope, a runaway train, a doomsday device, a real-life Frankenstein story complete with transplanted body parts, doddering creatures, and monstrous expense. The years bring fears, and we protect ourselves from the specter of death by throwing our wallets at it. When it comes to preserving me and mine, cost is no object.

Please understand that the medical industry won't let go of us until our life's potential (some would say ability to pay) has been exhausted. Meanwhile, religion and law stand on the sidelines cheering the sanctity of life, fully prepared to rush onto the playing field if the contest gets rough. To the upcoming generations who'll spend their earnings keeping old-timers alive as long as possible, I say: Good luck trying to live. And to those nearing life's end: Good luck trying to die.

The U.S. faces a seemingly unapproachable no-win situation, with life expectancy rising and oldsters living on at public expense. We don't have the political will (or the aesthetic inclination) to dress-down seniors, yet we must. Conventional mechanisms are paralyzed and we need a bold new idea, a maverick solution, which is why I thank Sarah Palin for envisioning the way out: death panels.

Call it reverse psychology, political jiu-jitsu, pity for a struggling commander-in-chief, or just plain nonpartisan patriotism, but Palin did her best to hand the blueprint over to Obama. The Governor and her camp, utilizing the power of suggestion, strove to shepherd the President in the right direction. That's the reason Palin ceded him ownership of her invention right from the start. "The America I know and love is not one in which my parents or my baby with Down Syndrome will have to stand in front of Obama's 'death panel'." Palin and her allies tried desperately to encourage Obama by giving him credit, before it had even occurred to him, for wanting to "pull the plug on grandma," and though Obama missed the point, not all within his administration turned away.

"Car Czar" sounds like a shill you'd see on late-night television, but it also happens to be the nickname of the lead auto industry advisor within the United States Treasury Department. In early 2009, Obama appointed Steven Rattner to that position so that Rattner could oversee the federal bailout of Chrysler and General Motors. Following his successful five-month tenure, Rattner returned to his regular job as a Wall Street financier—a guy who knows money. Here's how Rattner began an op-ed piece in the September 16, 2012 edition of *The New York Times*:

> Well, maybe not death panels, exactly, but unless we start allocating health care resources more prudently—rationing, by its proper name—the exploding cost of Medicare will swamp the federal budget. But in the pantheon of toxic issues—the famous "third rails" of American politics— none stands taller than overtly acknowledging that elderly Americans are not entitled to every conceivable medical procedure or pharmaceutical.

A few brave realists in the international community, facing similar circumstances, have also spoken out. Japan's Finance Minister, Taro Aso, in 2013 told his national council on social security reforms:

> I would wake up feeling increasingly bad knowing that [treatment] was all being paid for by the government. The problem won't be solved unless you let them hurry up and die.

The handwriting is on the nursing home wall (and it doesn't say "150 is the new 65"). Our inherent longing to live on, to self-perpetuate, will bankrupt the United States at the expense of young people and future generations. Medicine, technology, politics, law, and religion are on Granny's side, which leaves only one practical solution—Palin's. Considering what's at stake, I call it a modest proposal.

## PROOF THAT SELF DOES NOT EXIST

Some thinkers, including physicist Albert Einstein, philosophers David Hume and Derek Parfit, and psychologist Bruce Hood, challenge the notion of a separate self. Who am I to disagree with such luminaries, and in point of fact I can summarily prove that self does not exist.

TWO CLEAR PREMISES:

1. Every human self had another human self for a mother.
2. There have been a finite number of human selves.

ONE INEVITABLE CONCLUSION:

If even one self exists, it must have had a mother (premise #1), and that mother must have had a mother, and each a mother before that, resulting in an infinite back-chain of selves. But that would contradict our finiteness criterion (premise #2). Hence, if there were ever a single self to begin with, we would necessarily arrive at the logical impossibility of infinite selves.

Ipso facto ergo sum. Deus ex machina. Face the facts: it's over for that which you heretofore believed to be "you." Throw this book down, quit your job, move to Venice Beach and join the party. What have you got to lose? Trust me, I'm not making this up; I'm stealing it from philosopher David Sanford's 1975 exercise in faulty reasoning.

Once you've bought into premise #1 (mothers) you're sunk. The assertion presupposes that humanity sprawls forever backward in time, but fossils confirm that life goes back a mere 3.5 billion years, and *Homo sapiens* a paltry 200,000. What's certain, since evolution proceeds via miniscule increments, is that at some point a marginally-human being was born to a marginally pre-human mother. Further ancestral back-tracking would funnel us through the narrowing stages of life to its finite origin.

Don't let my fizzled syllogism sway you that self is what it seems. It isn't, and if you're not yet persuaded by the scientific evidence and my hitherto flawless reasoning, closing arguments in the case, upcoming momentarily, will seal the deal.

# A NEW MOI

*"Our minds are sort of electrochemical computers. Your thoughts construct patterns like scaffolding in your mind. You are really etching chemical patterns. In most cases, people get stuck in those patterns, just like grooves in a record, and they never get out of them."*
- Steve Jobs, Apple CEO

*"The true value of a human being can be found in the degree to which he has attained liberation from the self."*
- Albert Einstein

*"The burden of the self is lightened when I laugh at myself."*
- Rabindranath Tagore, poet

---

## CLOSING ARGUMENTS

*The Court*: Ladies and gentlemen of the jury, we will now hear closing arguments in the case of Science versus the Dominion of Self. The prosecution will begin.

*The Prosecution*: Thank you, Your Honor. Einstein called self an optical illusion of consciousness, a contention that the prosecution has backed up with evolutionary and experimental evidence. But let's be clear that an illusion isn't something nonexistent; it's an apparition that's not what it appears to be, and though an illusion may seem pleasant, unpleasant, or neutral, the ones that please us can be the most difficult to see through.

Self-understanding is like trying to observe, without a mirror, the rose-colored glasses that each of us wears. Others might tell us, "those Guccis are so *you*," but an impartial mirror—science—affords the only accurate reflection of self. The prosecution also contends that me-ness, when magnified to the collective level, becomes us-ness. Me and mine scales up to us versus them. To sum up the prosecution's case, self is supported by neither objectivity nor effect.

Let's imagine we're sitting in a theater watching the space movie *Gravity* in 3-D. I'll summarize the plot for those in the courtroom who haven't seen the film. Two astronauts, both great-looking, carom around like A-list billiard balls thanks to a tempest of orbiting debris and effects direction by Minnesota Fats. The female astronaut survives despite a lame and unnecessary backstory about her dead daughter.

As we sit there in our 3-D glasses, looking like a theater of aliens, pieces of space junk whiz alongside our heads and we duck. We buy into the illusion. We believe it even though we know we're watching lights flicker on a screen, even though we know the astronauts are actors, and even though we know the producers should have spent a little less on special effects and a little more on writers.

A movie exists but it's not what it seems. In actuality, cinema is an elaborate dramatization, built up from countless tiny elements such as film stock, boom mikes, body doubles, and gaffers. Scientific evidence indicates that individual identity is likewise a narrative, occasionally comical but too often serious, contrived by inborn bit players that have little or no awareness of their role in the final production—moi.

*The Court*: Thank you for the film class. Would the prosecution care to present any other general principles before reviewing the facts of the case?

*The Prosecution*: One more Your Honor, a short parable about our human situation that was told to me by Richard Dawkins and Daniel Dennett.

*The Court*: Proceed if you must.

*The Prosecution*: Let's suppose that each one of us is a wealthy adventurer, like Richard Branson or Ariana Huffington, and that we opt to live in the twenty-fifth century. We decide to build hibernation pods and enclose them in protective encasements, giant robots, but rather than preprogram our robots with a response to every possible situation, we embed decision-making modules that can assess incoming data and optimize strategy on the fly. In other words, to better protect ourselves we allow our robots a kind of self-control.

Your Honor that's humankind's situation, but we're not the ones in the pod; we're the lumbering robots. Our creators in the pod, that we're meant to preserve, are genes. It's difficult to see, because nature programmed us to focus on ourselves, that humanity's mission is to keep our genetic skippers alive. I ask the Court: Why else would we crave something as messy, dangerous, and psychologically bedeviling as sex? In animals with simple nervous systems genes keep behavior on a short leash, a stimulus-response scheme, but in human beings genes operate largely by remote control, via a long leash called mind, tethered to an insulating viewpoint called self.

*The Court*: Have you finished with the robot analogy?

*The Prosecution*: Yes, though it might prove entertaining if the metaphor were pushed to encompass robot reproduction.

*The Court*: Thank you for sparing us. Please recapitulate the facts of the case, starting with the cosmos.

*The Prosecution*: Cosmology informs us that you and I originated in the Big Bang, that each one of us is composed of literal star dust, that spacetime is immense beyond human comprehension, and that time, so critical in the development and maintenance of self, is more enigmatic than convention lets on.

*The Court*: Please review probability and randomness.

*The Prosecution*: Your Honor, our brains are poorly designed for probabilistic thinking, which is why it's hard to grasp that the unlikeliest events will eventually occur given exactly what the universe affords, vast opportunity. Each of us is a highly improbable but in no way exceptional eventuality. Professors Diaconis and Mosteller summed up the evidence as follows: "With a large enough sample, any outrageous thing is apt to happen."

*The Court*: Yes, the Michael Jackson case comes to mind. Please continue with the origin of life.

*The Prosecution*: Experts attested that life arose via molecular happenstance, then evolved via DNA replication and natural selection, which favors genetic traits that foster reproductive fitness. Matt Ridley testified that, "If genes induce their bodies to eat, have sex, rear children, and fend off rivals, then the genes themselves will be perpetuated."

*The Court*: And evolution?

*The Prosecution*: Whenever entities propagate in differential number based on expedient attributes, as in natural selection, some will preponderate while others fade. What else could happen? A conga line? Charles Darwin testified that, "The universe we observe has precisely the properties we would expect if there is, at bottom, no design, no purpose, no evil, no good, nothing but blind pitiless indifference."

*The Court*: Sounds cold, but we're here to face the facts. Please remind us about the brain.

*The Prosecution*: Your Honor, evolution benefitted creatures that could exert control over their environment, resulting in increasingly complex nervous systems and culminating in the human brain. Our brains developed over millions of years, with new capabilities lumped on top of old ones, which means that automatic systems such as those for aggression, sex, and selfishness continue to activate, even in situations where they're now counterproductive. Witnesses noted that almost all brain processes operate outside of consciousness, and neuroscientist V.S. Ramachandran testified that, "What each of us regards as his or her own intimate private self is simply the activity of these little specks of jelly in our heads, in our brains. There is nothing else."

*The Court*: And the mind?

*The Prosecution*: Mind, likewise, is fundamentally autonomous, a powerful elephant programmed by nature for quick decisions and self-furtherance. Consciousness, the rider, is more concerned with justifying the elephant's behavior than amending it. Evidence revealed that we're lazy, biased, and overconfident; that we judge people and experiences superficially; that we take credit for our bestowed attributes at the same time that we condemn other people for theirs; and that we elevate our needs and desires above others'.

Experts attested that we live in a brain-created virtual reality, one in which the unshakable omnipresence of words, thoughts, and concepts obscures the underlying irony that a thought-built world is a shadow of the real one. Lastly, we learned that personal identity is split between an experiencing self, which lives moment-to-moment, and a narrating self, which revisits the past and imagines the future. Philosopher David Hume testified that our minds are, "nothing but a heap or collection of different perceptions, united together by certain relations and supposed, though falsely, to be endowed with a perfect simplicity and identity."

*The Court*: Emotions?

*The Prosecution*: We discovered that feelings come and go more unguidedly even than thoughts, that happiness, the most sought after emotion, is largely beyond our control, and that natural selection has enhanced our fear reactions, which include anxiety and worry, to help safeguard our genetic heritage. Mark Twain testified that, "I have had a great many troubles, but most of them never happened." We learned that anger, another predominantly unproductive emotion, arms us for self-defense, and that we instinctually embrace our emotions, project their cause onto the environment, and justify them at others' expense. Even though feelings arise nonconsciously, once they inhabit us we take them to be ourselves and we act on their behalf. Oscar Wilde testified that, "The advantage of emotions is that they lead us astray."

*The Court*: And finally, please review the testimony regarding consciousness and self.

*The Prosecution*: Thank you, Your Honor. Consciousness permits us to include ourselves in a mental model of the world, and it allows us to run internal simulations in order to forecast outcomes. Self ensures that we benefit, directly or indirectly, from our activities. Several experts testified that the story of moi is just that, a historical fiction created by the mind, which transcribes the elephant's unconscious pursuits into a commendatory autobiography about the rider. Christopher Chabris and Daniel Simons encapsulated the prosecution's case: "Ultimately, seeing through the veils that distort how we perceive ourselves and the world will connect us, for perhaps the first time, with reality."

*The Court*: Thank you Counsel. You'll have an opportunity for a brief rebuttal after the defense has presented its case.

*The Defense*: Your Honor, I thank you. A number of philosophers and even a few scientists have spoken in defense of self's exceptionality. They've pointed to complex systems, to inherent irreducibility, and to "properties over and above the properties invoked by physics." But their testimony is unnecessary. The decision in this case is straightforward, because self is literally self-evident.

The prosecution has charged each of us in the courtroom with misdemeanor or felonious self-infatuation. We have listened to outside witnesses and heard scientific evidence, but the only genuine expert, the sole insider, never took the stand. I ask you, ladies and gentlemen of the jury, to at this very moment look within yourselves. Do you not find a home base—an identity, a vantage point, a viewport onto the world—comprised of memories, feelings, opinions, and a unique point of view? I ask next whether that singular perspective, your own self, dwells at the center of existence. Please look around. In whatever direction you turn, the universe lies before you. Your Honor, ladies and gentlemen of the jury, no amount of third-party evidence can supersede the eyewitness testimony of subjective experience. The defense rests.

*The Court*: Thank you for your brevity Counsel. I, like many no doubt, find your case rather appealing. Would the prosecution like to respond?

*The Prosecution*: Your Honor, what we've just heard is nothing more than Cypher's defense, straight out of *The Matrix*. "I know this steak doesn't exist. I know that when I put it in my mouth, the Matrix is telling my brain that it's juicy and delicious. After nine years, you know what I realize? Ignorance is bliss." The jury would be better advised to heed the words of Morpheus.

> Have you ever had a dream, Neo, that you were so sure was real? What if you were unable to wake from that dream? How would you know the difference between the dream world and the real world? You've been living in a dream world, Neo. Welcome to the real world.

Your Honor, we've worn our rose-colored glasses so long that we no longer realize we're wearing them. Our true selves are best reflected in the mirror of science. Evidence shows that each of us, physically and biologically, is the product of unwitting algorithms along with random chance. Plus, our thoughts, emotions, consciousness, and sense of self are byproducts of the brain, complex electrochemical alignments, refined by natural selection to enhance our reproductive prospects. In other words, selfness is an implanted sensation honed by evolution for genetic advantage. It's not supported by objectivity or predominantly beneficial.

Identity, after all, is an artifact of circumstance. There but for the grace of God go I. Ladies and gentlemen of the jury, if mad scientists were to replace each of your atoms, one by one, with those of Jennifer Lawrence you would become her, which is why, in closing, I remind you that my number is in the phonebook.

*The Court*: Counsel for the defense?

*The Defense*: Your Honor, we know what we know, and what we know best is our own individuality. Instead of poking around behind the curtain, let's pop open the champagne and celebrate self. What harm could it do? What could possibly go wrong?

## LEVELS OF CONSCIOUSNESS

According to Einstein, no problem can be solved from the same level of consciousness that created it. What that tells us is that Einstein acknowledged levels of consciousness and that he felt they were related to problem-solving. If we define consciousness as cognizance of what's going on, then I would be less conscious when I'm sleeping and more conscious than my cats, because while cats have keen senses, their self-understanding registers slim to none. For me, consciousness refers to cognizance of what's *really* going on, since evidence demonstrates that we're intrinsically prone to misconception, which means that upleveling consciousness involves edging closer to reality.

I'm not talking about spirituality, and I'm not talking about social or political awareness. Elevated consciousness, in my unadorned view, is no more than a prudential shift in focus. It's gaining control over our minds, attending to the actual instead of the reflexive. It's empowering System 2, the rider, to overlook the elephant's perpetual projections, and it's transitioning from the narrating self, which habitually elaborates past and future, to the experiencing self that abides in the moment.

Natural selection built up our brains one layer on another, never undoing the vestiges, and fashioned our minds for survival not wisdom.

As a consequence, we've wound up with three mental predispositions that now hold back consciousness: the like-o-meter[165], monkey mind, and psychological time.

The most important words in the elephant's vocabulary are like and dislike, or approach and avoid, because creaturely survival depends recurrently on precisely those assessments. An automatic like-o-meter runs continuously in each of our heads, as it does in every organism, passing split-second judgment on everything around us. Research shows that we have an immediate like/dislike reaction to virtually every event and object in our environment, though we're almost always unaware of our initial nonconscious response.

The like-o-meter binds consciousness because we instantaneously categorize every experience. We instinctively place it in one of three boxes—good, bad, or insignificant—and then, depending on the box, we respond according to innate and learned patterns. As we unthinkingly react, the rider rationalizes our response so that it "makes sense" to us, and so we can defend it to others.

If a perception is "good," we embrace it and try to freeze time right there; then when the thought or feeling invariably fades we strive to repeat the experience that evoked it. If a sensation or occurrence is "bad" we push it away; we resist our own lives and turn from pieces of ourselves. But most of what happens is inconsequential, so we ignore all that and get back to our primal preoccupation: pursuing and avoiding. The like-o-meter and its resultant categorization guarantee that we judge continually and discount 80% of the world around us. Then we wonder why joy is rare and life seems flat.

---

### The Time of Your Life

Have you ever asked yourself whether other people's day-to-day world seems more exhilarating to them than yours does to you? In the following quiz, give yourself one point for each checkmark, two if that would make it more exciting.

1. I find that time passes:
   ☐Slowly  ☐Like a glacier (prior to global warming)  ☐Time passes?

2. I'm reading this book because:
   ☐There was nothing on TV  ☐I thought it was porn  ☐Everyone told me not to

3. When waiting in a supermarket checkout line, I:

☐Scoff at other people's purchases    ☐Contemplate *Cosmo* cleavage
☐Try to guess what's going to delay the line

4.  The only thing worse than boredom is:
    ☐Being a bore   ☐Watching NASCAR on a smartphone   ☐Having
    to go out and do something

5.  Boredom is a form of:
    ☐Emotional void    ☐Cultural criticism    ☐Jack Daniel's sales pitch

I challenged you earlier to stop thinking for thirty seconds. The reason that's nearly impossible is monkey mind, the nonstop barrage of thoughts, the voice in our heads that has a life of its own. As Mozart put it, "When I feel well and in a good humor, or when I am taking a drive or walking after a good meal, or in the night when I cannot sleep, thoughts crowd into my mind as easily as you could wish." Much of our thinking, quite naturally, involves figuring out how to get what we want and avert what we don't. Spurious approach/avoid thoughts pop up, such as "that person over there looks cute," or "this food is too salty," or "I wonder if there's anything good on TV tonight," but even our focused thinking, like formulating a plan, usually revolves around prepping for good and circumventing bad.

Thoughts bubble up to consciousness whether we solicit them or not, but in either case the conscious self immediately assumes ownership of them. We presume that we govern our thoughts but it's another case of adverse possession, claiming personal proprietorship of impersonal property, since thoughts move in and move out on their own schedule. If we knew how to evict them we'd be able to stop thinking for thirty seconds.

Consciousness-raising is about taking control of our minds. In the process, the rider becomes increasingly able to direct attention and is less distracted by innate predilections. But mental clarity doesn't result from intellectual exercise, we need to train the elephant, and research shows that the most effective training regimen is meditation. But be forewarned that meditation is mysterious and mystical, as mysterious and mystical as push-ups or Zumba.

### Zumba Explained

Actually, what do I know about Zumba? With its occult-sounding name it could involve Aztec trance-dancing or sacrificial rites (but not human sacrifice, because killing off customers would be a second-rate business model). Here's what I found out by visiting the official Zumba website:

> Founded in 2001, Zumba Fitness is a global lifestyle brand that fuses fitness, entertainment and culture into an exhilarating dance-fitness sensation! Zumba exercise classes are "fitness-parties" that blend upbeat world rhythms with easy-to-follow choreography, for a total-body workout that feels like a celebration. We offer different types of Zumba classes, plus DVD workouts, original music collections, apparel and footwear, video games, interactive Fitness Concert™ events, a quarterly lifestyle magazine and more.

> Right below that blurb, big block letters proclaim that Zumba is practiced by 14 million people, in 185 countries, at over 140 thousand locations around the world. According to *Business Insider*, Zumba was named 2012 "Company of the Year" by Inc.com, and is currently the largest fitness brand in the world.

> You're thinking yet again that somebody should call the CDC. First it was zombies, then zimboes, and now we're in the midst of a Zumba invasion. But this one's a vast improvement, because Zumba dancers dress in tank tops not tatters, and instead of shambling they samba. Based on my true-life experience with salsa dancing I'm prepared to concede that Latin dance is a mystery, but I refuse to say the same for push-ups, because the only other creatures that perform them, male lizards, do so exclusively to attract females. And if that happens to sound like somebody you know, it's because the reptilian brain in a human operates exactly like a lizard's.

Psychological time is the third impediment to consciousness. The band Chicago famously asked us two questions: Does anybody really know what time it is? Does anybody really care? Science would answer no to the first question and yes to the second. Not only don't we know what time it is (even our most accurate instruments, the atomic clocks in Colorado and Paris, are not 100% accurate), we don't even know what time *is*. In 397 A.D., St. Augustine wrote, "It is in my own mind, then, that I measure time. I must not allow my mind to insist that time is something objective." The Persian philosopher Avicenna (980-1037 A.D.) agreed, declaring that, "Time is merely a feature of our memories and expectations," which, in a nutshell, is the opinion held by many scientists and philosophers today.

But does anybody really care? Indeed, we all do. Psychological time is our inborn penchant for rehashing the past and envisioning the future. As Pascal observed, "We never keep to the present. We anticipate the future as if we found it too slow in coming and were trying to hurry it up, or we recall the past as if to stay its too rapid flight. We are so unwise that we wander about in times that do not belong to us, and do not think of the only one that does."[166] In short, the narrating self has displaced the experiencing self. A veil of words, thoughts, and concepts has obscured life's inherent immediacy. Over the course of evolution our attention has drifted from the experiential to the conceptual, from the real to the imagined.

Without the past we'd have no identity, and without the future there'd be nothing to protect, which is the reason nature equipped us with a built-in time machine that helps solidify individual identity. We can reinforce our personal robustness (and that of our genetic skippers) by reconsidering triumphs and mistakes, and by reckoning ahead for self-perpetuation. If that were less true, we lumbering robots could be living it up in the present moment, Zumba dancing and practicing up for reproduction.

## MIND OVER CHATTER

Research shows that meditation, also called mindfulness training, is an efficient way to train the elephant. It dials down the like-o-meter, muffles the thought barrage, and defuses the time machine. Professor Mark Muesse, a graduate of Harvard, defines mindfulness as the skill of attending to one's experience as it unfolds without the superimposition of commentary or conceptualization, and meditation as a set of exercises that expand and refine mindfulness. For Dr. Muesse, meditation is an escape *into* reality. It isn't self-centered, it's self-ignorant, because not only does mindfulness practice help quell the notion of a static identity, it helps us feel comfortable without a fixed self.

Hundreds of scientific studies have shown that meditation is a simple, safe, and effective technique that helps expose and dislodge our instinctive but counterproductive mental habits. And for the record, while I make no claim to wisdom, I do practice formal meditation (by "watching" my breath) once or twice daily for 15 minutes, and I drag my attention (kicking and screaming) into the here and now whenever I can remember to do it.

According to the Mayo Clinic, emotional benefits of meditation include gaining a new perspective on stressful situations, focusing on the present, building stress-management skills, increasing self-awareness, and reducing negative emotions. As for physical health, the Mayo Clinic finds evidence that meditation can help with anxiety disorders, asthma, cancer, depression, heart disease, high blood pressure, pain, and sleep problems.[167] Meditation has also been shown to help the terminally ill accept death more peacefully.

To recap: Consciousness is the power to direct attention. We can raise consciousness by transferring mental governance from the elephant to the rider. Once we intellectually understand our predicament we're more likely to act, at which point meditation is the best way to train the elephant because it's safer, cheaper, and at least as effective as drugs or cognitive behavioral therapy.

Rebecca Gladding and Jeffrey M. Schwartz, both psychiatrists, have studied the neural effects of meditation. Gladding refers to the medial prefrontal cortex as the "Me Center" because it's a part of the brain that continually references back to our individual experiences and perspective, as when we think about the future, reflect on ourselves, or engage in social interactions. The lateral prefrontal cortex, however, is attuned to a more rational, logical, and balanced perspective. Gladding calls this the brain's "Assessment Center," designed to temper emotional responses, override automatic behaviors, and decrease the tendency to take things personally.

Studies show that our brains, before mindfulness practice, almost always exhibit strong neural connections between the Me Center and the regions for bodily sensations and fear, so whenever we feel anxious or experience a physical sensation we tend to assume there's a problem. For most of us, the medial prefrontal cortex, the Me Center, processes the bulk of our perceptual information. Neural over-reliance on the Me Center explains why we can get stuck in repetitive thought-loops about our lives, our mistakes, how people feel about us, our looks, and our health. The medial prefrontal cortex, in short, is one of self's staging grounds.

Our fear and bodily sensation circuitry is strongly connected to the Me Center, whereas Assessment Center circuitry is not, so it's easy to overvalue and overprotect ourselves. Research shows that meditation helps uncouple the Me Center from both the fear and bodily sensation centers, and strengthens connections between the Me Center and the Assessment Center. The outcome is decreased anxiety and increased equanimity. A fringe benefit, which Buddhists have long touted, is that mindfulness practice amplifies compassion for others. Good will grows because meditation enhances connections between the Me Center, the medial prefrontal cortex, and the insula, a part of the brain that helps us infer other people's state of mind and which fosters empathy. As Dr. Gladding points out, and as evidence attests, there's simply no substitute for meditation.

Oxford psychologist Derek Parfit maintains that personal identity makes self-interest seem more rationally compelling than any moral principle.[168] That is, it feels natural to take care of number one before helping others. According to Jonathan Haidt, the chief obstacle to world peace and social harmony is "naive realism"—taking ourselves at face value and blithely accepting intrinsic blind spots—because naive realism is easily ratcheted up from the individual to the collective level. "My group is right because we see things as they are."[169]

Consciousness-raising is about migrating mental governance from elephantine predispositions to the inner savant, attending to the world around us instead of to ourselves. It's not spirituality (or maybe it is, but that's irrelevant), it's how scientific enquiry suggests we move forward, especially if Einstein is correct that we won't solve our problems at the level of consciousness that created them.

## RELIGION = TRUTH, FUBAR

Buy into two, get one free. Having now endured my exposition of science and philosophy, feel free to roll your eyes as I elucidate religion as well. It turns out that religion isn't particularly difficult to fathom, because it boils down to consciousness: Krishna consciousness, the Buddha nature, Christ realization. It's about the inner experience of the original prophets and their attempts to share it with disciples. Sadly, the conveyance failed early and often, and we've ended up with the history of the church and today's organized religions.

Scripture consists of teachings designed to uplift consciousness, but followers can't always tell parable from reality. If you were writing an instruction manual that described how to change a tire, which do you suppose might work better: a) A father and his daughter traveled to town in order to buy flour for the bread she would bake. On the way back, their cart struck a stone and one of its wheels broke. The father propped up the cart using a fallen tree limb while the daughter replaced the broken wheel with a spare they carried in the cart. But the daughter, who was daydreaming of worldly matters, forgot to carefully pound in the wooden linchpins that affix the wheel to the axle. Just five hundred cubits down the road the wheel came loose, spilling the bags of flour and throwing father and daughter from the cart. The father, brushing flour from his overcoat, reprimanded his daughter, explaining that "she who dwells on the affairs of men instead of her duties, by fate shall never reach her home at the end of destiny's road," or b) Pay careful attention when tightening the lug nuts.

---

### The Story of Adam and Eve

The story of Adam and Eve offers another example. In a mere 199 words, Genesis 1:26 to 1:30, God made man in His likeness, and then woman, gave them dominion over plants and animals, and commanded them to reproduce. But that's only the trailer for the story of Adam and Eve, told in Genesis 2:4 to 3:24 and summarized below.

God created Adam and placed him in the Garden of Eden with only one restriction; that he not eat from the tree of the knowledge of good and evil. If Adam did he would die. God then created Eve to be Adam's companion. Adam and Eve were unashamed of their nakedness until a snake convinced Eve to eat the forbidden fruit, which she shared with Adam, whereupon they felt shame and covered up with fig leaves. Adam told God that Eve had given him the fruit, and Eve told God the snake had tricked her. As punishment, God made snakes and humans enemies, made childbirth painful, and made men toil for their food. God then expelled Adam and Eve from the Garden.

Rabbi Manis Friedman, writing in the *Huffington Post*, asks what we should make of this story.[170] We know from earlier in Genesis that God made man in His image, then instructed him to be fruitful and multiply, and we know that sin is doing what God said not to do or not doing what God said to do. Since Adam, Eve, and the Almighty were on good terms up until the apple incident, we can assume that the couple had been doing what God had told them to do, reproduce. Yet the Bible contains no reference to conception or birth until Genesis 4:1, after Adam and Eve's expulsion from Eden. In other words, there's no record of a single conception or birth in the Garden, even though we have to assume that Adam and Eve were following God's directive to multiply.

We're forced to conclude that Adam and Eve broke Eden's one and only rule posthaste. As soon as God made man and woman in His image, they violated His sole edict. That doesn't speak well for Adam and Eve, God's image, or by inference, the Almighty Himself. Now imagine you're the first man or woman. If God spoke directly to you and commanded you not to eat this one particular item, would you say, "I'll see what I can do"?

And what kind of father tells a child "Don't do X," then says, "because if you do, Y will happen"? That would be allowing for, if not insinuating, X. God forbids Adam to eat the fruit then adds, "for in the day that you eat of it you shall surely die." Why mention the day that you eat of it? The Lord might as well specify next Tuesday and ask Adam to mark it in his planner.

And what does Adam do when God confronts him? Adam blames Eve. Eve blames the snake. God had already warned Adam that he'd die if he ate the apple, but as if that weren't punishment enough, He serves up more. If God were in the NFL He'd be flagged for piling on. He makes enemies of humans and snakes, then saddles women with painful childbirth and men with toiling to eat. Adam is kicked out of Eden and will work for food. He's the ancestral homeless person.

Many fundamentalists take the Bible at its word (e.g., the snake really talks), which is too bad because they miss the point. Adam and Eve is actually an oft-interpreted allegory of the inner—it's about states of consciousness. My two cents pertains to the end of the story, where God says, "Behold, the man has become like one of Us, to know good and evil." In my interpretation (let the eye-rolling begin), Adam and Eve's "it's all good including our nakedness" consciousness (the Garden) becomes polluted when they find themselves promoted, via ignorance or their own gluttony, to judgeship. Their like-o-meters get amplified and they spawn human wrongheadedness, original sin. Adam and Eve, to me, is a Frankenstein story in which tarnished consciousness incarnates an imp named moi. Now, instead of abiding contentedly in the Garden, accepting creation as it presents itself, Adam and Eve have each become "like one of Us." In other words, they've deified individual authority, the self.

Darwin saw a universe without good or evil (only blind, pitiless indifference), and Shakespeare wrote, "There is nothing good or bad but thinking makes it so." There's a bumper sticker that sums up one of the lessons I take from Adam and Eve: Don't Believe Everything You Think. Another moral relates to the new demigod, personal primacy, because as soon as the couple set themselves up as judge and jury, unknowingly or purposefully, they had elevated themselves above the rest of existence. Psychologist William James said, "The altogether unique kind of interest which each human mind feels in those parts of creation which it can call me or mine may be a moral riddle, but it is a fundamental psychological fact." Self is a moral riddle in that we instinctually place ourselves ahead of others.  Problems ensue.

Hindus call our distorted perspective maya, the veil of delusion. Buddhists say that mind creates dukkha, a feeling of unsatisfactoriness. Christianity calls humankind's inherent subjective state original sin, but "sin," at least in the New Testament's initial Greek translation, meant to miss the mark, like an archer who misses the target.[71] My point is that a foremost tenet of religion, before it went FUBAR, was that our ordinary state of consciousness is out of sync with reality.

The founding prophets apparently rediscovered the Garden, but how?  The scriptures tell us they experienced oneness, the universe as a totality, a single unfolding process.  They saw the same mechanism at work within themselves as they saw beneath every leaf and stone.  With hindsight, "God" seems like a needlessly baggage-laden moniker, but had they agreed to call the phenomenon "Elvis," we'd now be debating Elvis's existence, which some of us do anyway.

George Eliot wrote, "Will not a tiny speck very close to our vision blot out the glory of the world, and leave only a margin by which to see? I know no speck so troublesome as self." The prophets could see clearly because they had removed the speck from their own eyes.  They had abandoned the narrating self's private life story and honed immediate awareness, the experiencing self.  The hone they almost certainly utilized was mindfulness meditation.

We know that's what Buddha did, and we know that Jesus was somewhere doing something between ages 12 and 30.  The Bible states only that during those eighteen years, "Jesus increased in wisdom and stature, and in favor with God and man." Thefreedictionary.com defines wisdom as "The ability to discern or judge what is true, right, or lasting; insight." Jesus could see clearly, but how did he become a sage?  We know it wasn't by watching Dr. Phil on television.  Surely Jesus had been cultivating his mind, perhaps via prayer—not the begging kind but the focusing kind—a traditional form of mindfulness practice.  "And when he was demanded of the Pharisees, when the kingdom of God should come, he answered them and said, The kingdom of God cometh not with observation.  Neither shall they say, Lo here! or, lo there! for, behold, the kingdom of God is within you."  It's a state of consciousness.

The original seers described their inner reality in the vernacular of the culture, to a troupe of disciples who endeavored to comprehend and chronicle the prophets' teachings. What followed was a millennia-long game of telephone. The intermediaries, almost none of whom had ever experienced the prophets' state of mind, passed along an increasingly distorted message, often as accurately as possible and sometimes with a personal twist, until today, the original teachings are buried in a pile of [check all that apply:    ☐Uplifting wisdom, such as "an eye for an eye" ☐Gluten-free Communion wafers    ☐Mormon marriage certificates ☐Boxes in the back of the Popemobile    ☐Sexual abuse settlement checks    ☐Horse hockey].

Organized religion spread thanks to idiomatic language, garbled transmissions, conceptual misinterpretation, and political spin. Greed and self-interest augmented the process. But wait, there's more; more than one game of telephone. Each prophet and a number of followers initiated a divinity dialathon, the end result of which has been thousands of off-course sects whose views differ amicably, spitefully, or violently. In 2012, the Center for the Study of Global Christianity at Gordon-Conwell Theological Seminary estimated that there are 43,000 Christian denominations worldwide, and of course they had no reason to address factions within other major religions. Take the Shiites and the Sunnis—please! To make a long story short, the Garden got overrun with weeds. The utter guilelessness of present-moment awareness and creaturely compassion went FUBAR.

The Lord doesn't work in mysterious ways. He doesn't work at all. He sits around throwing darts and rolling the bones. Einstein got this one wrong, since God does nothing but play dice with the universe. God is a name, contaminated beyond recovery, for the universal process, the ongoing molecular dance, the flux of purposeless happenstance from which each of us briefly materializes (He maketh us) and to which we shortly return (He taketh us back). There but for the grace of Chance go I. In truth, all of our prayers are answered, because whatever happens, that's the answer.

Scientists call the height of consciousness a "peak experience," a term coined by noted psychologist Abraham Maslow, who believed, as have others, that all religions are based on the insights of somebody's peak experience.[172] A 2006 double-blind study by neuroscientist Roland Griffiths and his colleagues at John Hopkins Medical School revealed that psilocybin could induce peak experiences in most subjects, and in the team's 14-month follow-up study, a majority of volunteers reported an overwhelmingly positive experience, specifying that it ranked among the five most personally meaningful events of their lives.[173] Griffiths was following on the legendary work of Harvard psychologists Timothy Leary and Richard Alpert, the latter of whom went on to become Ram Dass.

Maslow referred to an enduring peak experience, such as that of the original prophets, as self-actualization. In some traditions that state of mind is called enlightenment, in others liberation, but I call it the 100% solution because the transformation of consciousness is complete. Identity has fully migrated to the experiencing self. The narrating self, whose normal job it is to spin out a personal life story, atrophies from disuse, thus liberating us from our selves. Innate self-interest finds no footing as attention shifts from moi to existence as it unfolds.

For me, of course, this is all intellectual supposition. I'm as close to enlightenment as I am to winning Wimbledon. I'll pursue neither. To begin with, the 100% solution is nearly unachievable. Second, the effort can devour a lifetime of energy. Third, the history of religion shows that it's too easy to veer off track. And fourth, I forgot where I put my tennis racquet. Instead, I advocate the 7% solution, a concept I introduced in a three-page short story entitled "The Liberal Uprising of 2024," and which I'll describe in more detail shortly.

## The Liberal Uprising of 2024

Dr. Hans Decker closed his eyes and opened them. The assessment room was silent, except for an occasional soft click from one of its older biosensors.

Decker continued: "Ms. Miller, Rachel, please answer the question. Why did you slit your wrists?" His gaze shifted from the BrainScan3000 to her clenched face.

"Madonna took my gun away in 2025. All I had was a kitchen knife."

"President Madonna took everybody's guns," Decker replied. "What I mean is, why did you want to kill yourself in the first place?" Most of the information for his report would stream directly from the scanners, but intentionality laws required him to ask.

In the two decades preceding the Liberal Uprising of 2024, science had proven that ninety-three percent or more of human behavior is beyond conscious control, the product of nature (genes), nurture (upbringing), and brain damage caused by exposure to Fox News and MSNBC. The remaining conduct, at most seven percent, was where any vestige of personal responsibility might reside. It was Decker's job to root it out, then file a disposition report along with a recommendation.

Genuine "seven-percenters" were unpredictable because they possessed a modicum of independent thought. Under special circumstances they could be held accountable for their actions, and that's where Decker came in. In Decker's experience, though, most people turned out to be "zero-percenters," knee-jerkers to the core.

Rachel lowered her eyes. "My husband divorced me and ran off with our golden retriever. They got married."

"That's legal now," Decker reminded her. Decker had supported The Interspecies Marriage Act of 2026. Love is love, he felt.

"I know," Rachel said, "but it's not the first time he cheated on me with a nonhuman."

Decker had a pretty good idea what was coming next, because he'd seen it before. He changed the subject. "We'll get back to that. But why didn't you just check into an assisted dying facility—there are three in town—instead of slitting your wrists?" (Directly after the revolution, the Right to Life movement had dissolved, replaced by the Right to Death industry.)

Rachel looked up at him. "I live outside town and my bicycle is broken. It was too far to walk."

Decker regarded her skeptically. She was wearing a skintight red body suit, which Decker could see fit her rather well. (That wouldn't go in his report.) He noted the Under Armour logo, which to Decker still looked like an "H." The brand was so popular now that Under Armour had become the world's second largest corporation, right behind Apple. Rachel looked fit enough to walk into town.

Rachel softened when she noticed Decker eyeing her. "I don't like leaving a mess. If I still had a car, I would have used a facility in town."

As soon as liberals seized power in 2024 and installed Madonna, she authorized a fourth federal bailout of the U.S. auto industry, plus legislation abolishing fossil fuel-powered vehicles. Electric cars of all kinds immediately flooded the market. In November 2025, Ford introduced an electric monster truck, the FU-150, which was the best-selling vehicle in Texas. (A year later Texas became a separate country, because when it threatened to secede, President Madonna promptly signed the secession papers.)

The electric car era came to an end in early 2027 when authorities discovered that over a thousand teenagers had died after snorting lithium extracted from car batteries. (Lithium was one of three substances not covered under The Drug Legalization Act of 2025.) The progressive majority, already reveling in its newfound rule, raised their Chardonnay glasses once again as the U.S. became a pedal-powered nation. Horses too were permitted, as long as riders carried a pooper-scooper.

"Maybe," Decker thought to himself as he and Rachel appraised each other, "she's a seven-percenter." He moved on: "Let's get back to your husband. You say he'd cheated on you earlier?"

"It's a long story," she replied.

Decker lightly held her gaze and thought to himself, "Yeah, but I bet I've heard it before." "Go on," he said.

Rachel propped herself up in the chair. "Do you remember when Miley Cyrus hosted the Academy Awards in 2024, right after the rebellion?"

"I missed it."

"Well, every year they have this segment called 'In Memoriam.' They pay tribute to a list of film celebrities who died that year."

"I've seen them do that, but not in 2024," Decker explained.

"Unfortunately, I *did* see it, and so did my husband. Miley Cyrus had them wheel out the casket of each dead celebrity, one after another, and right there on center stage she opens the casket and *has sex* with the dead celebrity! Man or woman. She just jumps in and starts grinding all over them."

"That's legal now," Decker reminded her, although he himself had not supported The Sexual Freedoms Act of 2024. In Decker's medical opinion, necrophilia was not a sexual orientation.

"My husband loved it, and he told me it was the highest-rated Academy Awards ever. People actually wanted the show to run *longer*. But The Oscars wasn't the problem."

Decker suspected what she'd say next. "Go on."

"Well, the show turned out to be so popular that they came out with these life-size electric Miley Cyrus dolls. You must have seen one, the Twerk-a-Magic 5000, by Hasbro. Lots of guys have them, and lots of women too."

Decker cut in: "Slow down, let me guess." He regarded Rachel. Ten minutes ago, hunched into her chair, she'd seemed fragile, but now that she was upright and animated, Rachel radiated potency. "I'll bet your relationship went downhill from there, right?"

Rachel leaned back, exhaled. "Right, I suck at twerking."

Decker doubted that. His job, though, was to evaluate Rachel, not fantasize about her. He needed to be especially careful because in 2026 the definition of sexual harassment had been revised to, "Any sexual thought or action involving another being without his, her, or its express written consent."

Following the revolution, healthcare finally progressed. Congress implemented a single-payer system, the payer to be named later. (They eventually named Donald Trump as payer, because he could afford it and because he had made us all sick for so long.) The program, dubbed MadonnaCare, provided health care to all Americans and their house pets (but not to livestock, which upset animal rights activists). A citizen could be turned down for only three reasons: eating processed foods, being out of shape, or drinking Budweiser.

Decker fumbled for the array of switches on his desk. He flicked off the BrainScan3000 and the other biosensors. The assessment was over but no report would be filed. Rachel was a seven-percenter, he was certain. A seven-percenter was rare, especially an attractive one in a skin-tight Under Armour body suit. Her husband had left her for a pooch; she deserved better. He looked in her eyes and said, "Let's get of here. We can go to my place."

Rachel leaned forward, tilted her head ever-so-slightly to the side, and holding Decker's gaze replied, "OK."

They rode silently to Decker's house, he peddling fervently and she on the handlebars. They turned into his driveway, Rachel jumped off, and Decker laid the bicycle on the asphalt. As they hurried through the door, Decker tugged off his blazer while Rachel tidied her body suit and fluffed her hair.

They scuttled past the granite-top kitchen island, past the wine rack, past the yoga mats, past the Native American pottery, past the distressed-wood coffee table with the copy of *An Inconvenient Truth* on it, past the treadmill, and past the poster of Bob Marley hanging in the hallway. Decker hop-stepped into the bedroom, Rachel in tow.

Rachel gasped and her eyes bulged. On Decker's Japanese-style bed lay a golden retriever and a Twerk-a-Magic 5000. The doll had a paper cut-out of Madonna's face pasted onto it.

## THE 7% SOLUTION

For Socrates, the unexamined life is not worth living. He fails to mention the handicap of performing said examination through rose-colored coke bottle glasses. Given that mind operates largely outside of consciousness, and that ego defends itself, how can partisan examiners come to know and cultivate their authentic selves? Our predicament is akin to that of the seafarers on Neurath's boat: "We are like sailors who on the open sea must reconstruct their ship but are never able to start afresh from the bottom. Where a beam is taken away a new one must at once be put there, and for this the rest of the ship is used as support. In this way, by using the old beams and driftwood the ship can be shaped entirely anew, but only by gradual reconstruction."[174]

Our best bet, in other words, is to find and stand on our truest 7%, the aspect that we most trust and over which we hold sway, as we gradually expose and renovate its instinctual foundation. Consciousness is an artifact of the brain, and the 7% solution provides a bootstrapping mechanism, a recurrent feedback loop, in which conscientious thoughts and actions recursively upgrade our neural circuitry. Nobody intends to be a zero-percenter, an evolutionary or cultural knee-jerker, but nature fashioned us to follow our instincts and to fit in with society. It takes knowledge, incentive, and effort to rewire our preprogrammed selves and uplevel consciousness.

As we sit in the movie theater watching *Gravity*, and pieces of space junk whiz alongside our heads, we buy into the illusion. But not for the rest of our lives. When we exit the theater it was only a show. Psychiatrist David Galin states that, "it takes arduous training to modify or overcome the natural state of experiencing the self as persisting and unchanging." But if individuality is not what it seems, as science attests, then we can deal with the facts or live in the Matrix. Only a clear-eyed and resolute biochemical puppet can begin to pull its own strings.

The story of moi is yet another Frankenstein story. It's about creatures (you and me) stitched blindly together by evolution, given a jolt of power, and set loose on the countryside. The 7% solution is how the creature redeems itself.

Personal identity is a sensation that genes implant in us because it heightens our self-centeredness. A cat will defend its territory, but it won't sit around auditing accomplishments and charting next moves. When Oxford psychologist Derek Parfit ceased to fixate on a separate self, he wrote, "My life seemed like a glass tunnel, through which I was moving faster every year, and at the end of which there was darkness. When I changed my view, the walls of my glass tunnel disappeared. I now live in the open air. There is still a difference between my life and the lives of other people. But the difference is less. Other people are closer. I am less concerned about my life, and more concerned about the lives of others."[175] The 7% solution is a bottom-up plan that dovetails with top-down humanist agendas.

Self-devaluation matters, because what we consider "the problem" is often just a symptom of an underlying condition. One example is "the tragedy of the commons," economist Garrett Hardin's description of how individuals, acting rationally according to self-interest, deplete a shared resource. Environmental sustainability should be a no-brainer, because even the greediest exploiters have children and grandchildren, but at our current level of consciousness it's apparently a no-wayer.

Another of self's repercussions is income inequality. Right now, America's bottom 80% controls only 7% of the nation's wealth, which is not the 7% solution I have in mind. According to a 2014 NASA-funded study, the two most critical factors in societal disintegration are overuse of natural resources and wealth disparity. The research team concluded: "Collapse can be avoided and population can reach equilibrium if the per capita rate of depletion of nature is reduced to a sustainable level, and if resources are distributed in a reasonably equitable fashion."[76] Looking out for number one could land all of us in a pile of number two.

If personal identity is a gene-induced sensation, and existence a beauty contest where bestowed attributes and happenstance dictate our fate, why expend the energy to pull back life's curtain? Here's Richard Dawkins's answer: "After sleeping through a hundred million centuries we have finally opened our eyes on a sumptuous planet, sparkling with color, bountiful with life. Within decades we must close our eyes again. Isn't it a noble, an enlightened way of spending our brief time in the sun, to work at understanding the universe and how we have come to wake up in it?" Indeed, understanding is Part A of the 7% solution. Part B is shipbuilding.

You and I are bit players on the cosmic stage. Is it all sound and fury signifying nothing? Maybe Adam and Eve could make that call with their divine-like discernment, but it's above my pagan pay grade. I say the most prudent approach is to lighten up, dial back moi, and enjoy the show. False: I'm a dancer and my life is my dance. True: Life is the dancer, and it wants me to Zumba.

# NOTES

[1] Parfit, Derek.  Reasons and Persons.  1986.  Oxford Paperbacks.  P.281.
[2] http://www.treehugger.com/natural-sciences/the-pacific-garbage-patch-may-not-be-emtwiceem-the-size-of-texas.html
[3] Person of the Year: 75th Anniversary Celebration (Special Collector's Edition ed.). New York: Time Books. 2002. OCLC 52817840, reference from Wikipedia.
[4] Ken Burns documentary "The National Parks: America's Best Idea," 2009.
[5] http://scienceblogs.com/mixingmemory/2006/12/coolest experiment_ever.php
[6] Cox, Brian, and Cohen, Andrew.  Wonders of the Universe.  2011. HarperCollins.  p. 128.
[7] http://imagine.gsfc.nasa.gov/docs/science/know_l1/pulsars.html
[8] Cox, Brian, and Cohen, Andrew.  Wonders of the Universe.  2011. HarperCollins.  Back cover.
[9] Schultz, William Todd (2005).  Handbook of Psychobiography.  Oxford University Press.  p. 93.
[10] Elms, Alan C. (July–September 1977). ""The Three Bears": Four Interpretations". The Journal of American Folklore 90 (357).
[11] Tatar, Maria (2002). The Annotated Classic Fairy Tales. W.W. Norton & Company. ISBN 0-393-05163-3. P. 246.
[12] Cox, Brian, and Cohen, Andrew.  Wonders of the Universe.  2011. HarperCollins. p. 150.
[13] Cox, Brian, and Cohen, Andrew.  Wonders of the Universe.  2011. HarperCollins. p. 193.
[14] Scientific American Editors (2012-11-30). A Question of Time: The Ultimate Paradox (Kindle Locations 182-185). Scientific American. Kindle Edition.
[15] Arthur Eddington, The Nature of the Physical World, 1928, from Wikipedia.
[16] http://www.nytimes.com/2004/10/10/movies/10kapl.html? pagewanted= 1&_r=0 October 10, 2004 article by Fred Kaplan.

[17] http://science.nasa.gov/astrophysics/focus-areas/what-is-dark-energy/

[18] http://www.nobelprize.org/nobel_prizes/physics/laureates/2011/press.html

[19] http://science.nasa.gov/astrophysics/focus-areas/what-is-dark-energy/

[20] Jillette, Penn (2011-08-16). God, No! (Kindle Locations 932-933). Simon & Schuster, Inc.. Kindle Edition.

[21] Cassells, Schoenberger, and Grayboys, 1978, cited in Randomness, Deborah J. Bennett, 1998.

[22] Randomness, Deborah J. Bennett, 1998.

[23] Jonah Lehrer. How We Decide (Kindle Locations 791-793). Kindle Edition.

[24] http://www.urbanette.com/taken-for-granted/

[25] Norton Starr (1997). "Nonrandom Risk: The 1970 Draft Lottery". Journal of Statistics Education 5.2 (1997), cited in Wikipedia

26 http://www.theepochtimes.com/n3/787114-scientists-calculate-the-probability-of-your-existence/

[27] https://www.apologeticspress.org/apcontent.aspx?category=9&article=472

[28] Atkins, Peter (2011-03-17). On Being : A scientist's exploration of the great questions of existence (pp. 19-20). Oxford University Press. Kindle Edition.

[29] http://www.youtube.com/watch?v=3p47bGlZuDA

[30] Terminator 2: Judgment Day, 1991.

[31] Virus. Wolfhard Weidel. Translated from the German by Lotte Streisinger. University of Michigan Press, Ann Arbor, 1959, cited on http://library.thinkquest.org/C003763/index.php?page=origin06

[32] http://en.wikipedia.org/wiki/Abiogenesis#.22Primordial_soup.22_theory

[33] The Selfish Gene. Richard Dawkins. 1976. Kindle Edition.

[34] The Selfish Gene. Richard Dawkins. 1976. Kindle Edition.

[35] Ridley, Matt. The Red Queen: Sex and the Evolution of Human Nature (Kindle Locations 167-170). HarperCollins. Kindle Edition.

[36] http://en.wikipedia.org/wiki/Ray_Kurzweil

[37] http://en.wikipedia.org/wiki/Turritopsis_nutricula

[38] http://liberator.net/articles/SloanGary/Shaw.html

[39] Daniel Dennett, 1995, Darwin's Dangerous Idea, cited in The Robot's Rebellion: Finding Meaning in the Age of Darwin, Keith E. Stanovich, 2005.

[40] The Robot's Rebellion: Finding Meaning in the Age of Darwin, Keith E. Stanovich, 2005.

[41] Letter 2743 — Darwin, C. R. to Gray, Asa, 3 April (1860) from http://en.wikipedia.org/wiki/The_Descent_of_Man,_and_Selection_in _Relation_to_Sex#cite_note-5

[42] Evolutionary Explanations of Human Behavior, John Cartwright, 2001, p. 28.

[43] http://en.wikipedia.org/wiki/Physical_attractiveness#Facial_ attractiveness

[44] Steven Pinker The stuff of thought: language as a window into human nature page 371.

[45] Matt Ridley, from his book Genome, excerpted in The Oxford Book of Modern Science Writing, 2008.

[46] Atkins, Peter. On Being: A Scientist's Exploration of the Great Questions of Existence, p.47 (Kindle Edition).

[47] http://jcdverha.home.xs4all.nl/scijokes/4_6.html

[48] Richard Lynn. DYSGENICS: Genetic Deterioration in Modern Populations. Second Revised Edition. Ulster Institute for Social Research, 1930.

[49] http://en.wikipedia.org/wiki/For_Dummies

[50] http://www.livescience.com/699-painful-realities-hyena-sex.html

[51] http://pewresearch.org/pubs/303/gauging-family-intimacy

[52] Doug Erwin, from Fred Guterl, The Fate of the Species: Why the Human Race May Cause its Own Extinction and How We Can Stop It, 2012, Kindle edition Location 863.

[53] http://www.classroomtools.com/scilit.htm

[54] http://www.huffingtonpost.com/2008/09/28/palin-claimed-dinosaurs-a_n_130012.html

[55] http://creationmuseum.org/

[56] Soren Kierkegaard, writing as *Johannes Climacus,* Concluding Unscientific Postscript, 1846 p.317, from http://sorenkierkegaard.org/johannes-climacus.html.

[57] http://en.wikipedia.org/wiki/Evil_demon

[58] The Matrix, 1999, from IMDB.com

[59] The Man with Two Brains, 1983, from IMDB.com

[60] The Man with Two Brains, 1983, from IMDB.com

[61] http://www.pbs.org/wgbh/nova/body/brain-transplants.html

[62] V.S. Ramachandran, A Brief Tour of Human Consciousness, 2004, p.2.

[63] Francis Jacob, from David J. Linden, The Accidental Mind, 2007, p.6.

[64] Ebraheim NA, Lu J, Skie M, Heck BE, Yeasting RA, from http://www.ncbi.nlm.nih.gov/pubmed/9399453

[65] Mlodinow, Leonard (2012-04-24). Subliminal: How Your Unconscious Mind Rules Your Behavior (p. 142). Knopf Doubleday Publishing Group. Kindle Edition.

[66] Neil deGrasse Tyson.  Presentation at the Beyond Belief Conference, 2006.

[67] David J. Linden, The Accidental Mind, 2007, p.144.

[68] Wenda R. Trevathan, Human Birth, (New York: Aldine de Gruyter, 1987), p. 109

[69] Paul Medina, Brain Rules, Location 475 Kindle Edition.

[70] http://www.amazon.com/s/ref=nb_sb_noss_1?url=search-alias%3Dhpc&field-keywords=magnetic+bracelet

[71] Bruce Hood, The Self Illusion,p19.

[72] V.S. Ramachandran, A Brief Tour of Human Consciousness, 2004, p.3.

[73] David J. Linden, The Accidental Mind, 2007, p.33.

[74] Laura Fitzpatrick, Time Magazine, A Brief History of Antidepressants, Jan. 07, 2010.

[75] http://news.yahoo.com/stress-study-offers-clues-antidepressant-drugs-191030790.html

[76] Joseph LeDoux, Synaptic Self, 2002, p.297.

[77] http://en.wikipedia.org/wiki/Functional_magnetic_resonance_ imaging

[78] http://en.wikipedia.org/wiki/Functional_magnetic_resonance_ imaging

[79] https://ww4.aievolution.com/hbm1201/index.cfm?do=abs.view Abs&abs=5553

[80] Siegel, Daniel J., Pocket Guide to Interpersonal Neurobiology: An Integrative Handbook of the Mind (Norton Series on Interpersonal Neurobiology) (p. 4). Norton. Kindle Edition.

[81] Chabris, Christopher; Simons, Daniel . The Invisible Gorilla: And Other Ways Our Intuitions Deceive Us . Crown Publishing Group. Kindle Edition.

[82] Bedau, Mark A, in J. Tomberlin, ed., Philosophical Perspectives: Mind, Causation, and World. Vol. 11 (Malden, MA: Blackwell, 1997), pp. 375-399.

[83] Siegel, Daniel J. (2012-04-02). Pocket Guide to Interpersonal Neurobiology: An Integrative Handbook of the Mind (Norton Series on Interpersonal Neurobiology) (p. 112). Norton. Kindle Edition.

[84] LeDoux, Joseph (2003-01-28). Synaptic Self: How Our Brains Become Who We Are (Kindle Location 336). Penguin Group. Kindle Edition.

[85] http://www.mayoclinic.com/health/traumatic-brain-injury/DS00552/DSECTION=complications

[86] David J. Linden, The Accidental Mind, 2007, p.83.

[87] Hood, Bruce (2012-04-25). The Self Illusion: How the Social Brain Creates Identity . Oxford University Press. Kindle Edition.

[88] Dennett, Daniel C. (2013-05-06). Intuition Pumps And Other Tools for Thinking (p. 36). W. W. Norton & Company. Kindle Edition.

[89] Dawkins, Richard, Editor. The Oxford Book of Modern Scientific Writing. Oxford University Press. 2008.

[90] Mlodinow, Leonard (2012-04-24). Subliminal: How Your Unconscious Mind Rules Your Behavior (p. 130). Knopf Doubleday Publishing Group. Kindle Edition.

[91] Cartwright, John H. (2007-03-16). Evolutionary Explanations of Human Behaviour (Routledge Modular Psychology) (p. 24). Taylor & Francis. Kindle Edition.

[92] Plato. Plato in Twelve Volumes, Vol. 9 translated by Harold N. Fowler. Cambridge, MA, Harvard University Press; London, William Heinemann Ltd. 1925.

[93] Haidt, Jonathan (2012-03-13). The Righteous Mind: Why Good People Are Divided by Politics and Religion (Kindle Locations 101-104). Knopf Doubleday Publishing Group. Kindle Edition.

[94] http://www.dailymail.co.uk/news/article-1192674/I-banished-stockroom-says-disabled-shop-girl-suing-Abercrombie--Fitch-discrimination.html

[95] http://www.hofstra.edu/pdf/orsp_shahani-denning_spring03.pdf

[96] V.S. Ramachandran, A Brief Tour of Human Consciousness, 2004, p.76.

[97] Feynman, Richard. Surely You're Joking Mr. Feynman, W.W. Norton and Company, Inc. 1985, p 20.

[98] Seinfeld, The Wallet, 1992, at http://www.seinfeldscripts.com/TheWallet.html

[99] http://en.wikipedia.org/wiki/Mirror_neuron

[100] Jonah Lehrer. How We Decide (Kindle Locations 2107-2108). Kindle Edition.

[101] Hood, Bruce (2012-04-25). The Self Illusion: How the Social Brain Creates Identity (p. 65). Oxford University Press. Kindle Edition.

[102] http://www.brainyquote.com/quotes/authors/p/patti_stanger.html

[103] http://www.wired.com/science/discoveries/news/2008/04/mind_decision

[104] Jochem W. Rieger, Sebastian Baecke, Ralf Lützkendorf, Charles Müller, Daniela Adolf, Johannes Bernarding. October 07, 2011DOI: 10.1371/journal.pone.0025304

[105] Stefan Bode, Anna Hanxi He ¶, Chun Siong Soon, Robert Trampel, Robert Turner, John-Dylan Haynes. www.plosone.org/article/info%3Adoi%2F10.13712 Fjournal.pone.0021612. Published: June 27, 2011DOI: 10.1371/journal.pone.0021612.

[106] Haidt, Jonathan (2012-03-13). The Righteous Mind: Why Good People Are Divided by Politics and Religion (Kindle Locations 1633-1639). Knopf Doubleday Publishing Group. Kindle Edition.

[107] Haidt, Jonathan (2012-03-13). The Righteous Mind: Why Good People Are Divided by Politics and Religion (Kindle Locations 1482-1485). Knopf Doubleday Publishing Group. Kindle Edition.

[108] Haidt, Jonathan (2012-03-13). The Righteous Mind: Why Good People Are Divided by Politics and Religion (Kindle Locations 835-837). Knopf Doubleday Publishing Group. Kindle Edition.

[109] Hood, Bruce (2012-04-25). The Self Illusion: How the Social Brain Creates Identity (p. 82). Oxford University Press. Kindle Edition.

[110] Mlodinow, Leonard (2012-04-24). Subliminal: How Your Unconscious Mind Rules Your Behavior (p. 214). Knopf Doubleday Publishing Group. Kindle Edition.

[111] http://en.wikipedia.org/wiki/False_memory_syndrome

[112] http://en.wikipedia.org/wiki/Agent_detection

[113] http://www.equip.org/articles/does-religion-originate-in-the-brain/#christian-books-2

[114] http://genealogyreligion.net/hyperactive-agency-detection-devices-and-horny-antelopes

[115] Hood, Bruce (2012-04-25). The Self Illusion: How the Social Brain Creates Identity (pp. 119-120). Oxford University Press. Kindle Edition.

[116] http://en.wikipedia.org/wiki/Charles_Whitman

[117] http://en.wikipedia.org/wiki/Charles_Whitman

[118] http://en.wikipedia.org/wiki/Charles_Whitman

[119] News of the Weird, in Funny Times, March 2014.

[120] Dennett, Daniel C. (2013-05-06). Intuition Pumps And Other Tools for Thinking (p. 360). W. W. Norton & Company. Kindle Edition.

[121] http://en.wikipedia.org/wiki/Emotion_in_animals

[122] http://en.wikipedia.org/wiki/Paul_Ekman

[123] Ekman, Paul. Emotions Revealed. Holt Paperbacks, 2003, p.19.

[124] Mlodinow, Leonard (2012-04-24). Subliminal: How Your Unconscious Mind Rules Your Behavior (pp. 179-180). Knopf Doubleday Publishing Group. Kindle Edition.

[125] de Becker, Gavin (2010-01-20). The Gift of Fear (Kindle Locations 495-498). Gavin de Becker. Kindle Edition.

[126] de Becker, Gavin (2010-01-20). The Gift of Fear (Kindle Locations 4690-4692). Gavin de Becker. Kindle Edition.

[127] de Becker, Gavin (2010-01-20). The Gift of Fear (Kindle Locations 4838-4862). Gavin de Becker. Kindle Edition.

[128] Ekman, Paul. Emotions Revealed. Holt Paperbacks, 2003, p.110.

[129] de Becker, Gavin (2010-01-20). The Gift of Fear (Kindle Locations 203-205). Gavin de Becker. Kindle Edition.

[130] Ekman, Paul. Emotions Revealed. Holt Paperbacks, 2003, p.120.

[131] http://www.reuters.com/article/2013/11/12/us-arab-women-factbox-idUSBRE9AB00I20131112

[132] http://well.blogs.nytimes.com/2010/08/11/phys-ed-can-exercise-moderate-anger/?_php=true&_type=blogs&_r=0

[133] http://en.wikipedia.org/wiki/Hustler_Magazine_v._Falwell#cite_note-6

[134] http://www.imdb.com/title/tt0070707/trivia?tab=qt&ref_=tt_trv_qu

[135] http://thehowofhappiness.com/

[136] Lyubomirsky, Sonja (2007-12-27). The How of Happiness: A New Approach to Getting the Life You Want (pp. 41-42). Penguin Group. Kindle Edition.

[137] Jonathan Haidt. The Happiness Hypothesis: Finding Modern Truth in Ancient Wisdom (Kindle Locations 499-511). Kindle Edition.

[138] Jonathan Haidt. The Happiness Hypothesis: Finding Modern Truth in Ancient Wisdom (Kindle Locations 1147-1159). Kindle Edition.

[139] http://en.wikipedia.org/wiki/Laughter_in_animals

[140] Curtis, Valerie. Don't Look, Don't Touch, Don't Eat: The Science Behind Revulsion (Kindle Locations 42-45). University of Chicago Press. Kindle Edition.

[141] http://in.reuters.com/article/2014/02/20/australia-marsupial-discovery-antechinus-idINDEEA1J08020140220

[142] Tennov, Dorothy. Love and Limerence (Kindle Locations 142-146). Rowman & Littlefield. Kindle Edition.

[143] David J. Linden, The Accidental Mind, 2007, p.145.

[144] Roach, Mary (2009-04-06). Bonk: The Curious Coupling of Science and Sex (pp. 23-24). W. W. Norton & Company. Kindle Edition.

[145] Roach, Mary (2009-04-06). Bonk: The Curious Coupling of Science and Sex (p. 218). W. W. Norton & Company. Kindle Edition.

[146] http://www.misspiggyfans.com/Quotes/

[147] Rogers, Carl (1961). On becoming a person: A therapist's view of psychotherapy. London: Constable. ISBN 1-84529-057-7.

[148] Bernard J. Baars, editor. Essential Sources in the Scientific Study of Consciousness. Bradford Books. 2003 p.8.

[149] http://en.wikipedia.org/wiki/Hard_problem_of_consciousness

[150] Humphrey, Nicholas (2012-08-25). A History of the Mind: Evolution and the Birth of Consciousness (Kindle Location 1069). . Kindle Edition.

[151] Bernard J. Baars, editor. Essential Sources in the Scientific Study of Consciousness. Bradford Books. 2003 p.48.

[152] Humphrey, Nicholas (2011-01-31). Soul Dust: The Magic of Consciousness (p. 66). Princeton University Press. Kindle Edition.

[153] Humphrey, Nicholas (2011-01-31). Soul Dust: The Magic of Consciousness (p. 149). Princeton University Press. Kindle Edition.

[154] Chalmers, David J. Facing Up to the Problem of Consciousness. Journal of Consciousness Studies 2(3). 1995, pp. 200-219.

[155] Nunez, Paul L. Brain, Mind, and the Structure of Reality. Oxford University Press. 2010. p.251.

[156] Humphrey, Nicholas (2012-08-25). A History of the Mind: Evolution and the Birth of Consciousness (Kindle Locations 3056-3057). . Kindle Edition.

[157] Jonathan Haidt. The Happiness Hypothesis: Finding Modern Truth in Ancient Wisdom (Kindle Locations 1872-1874). Kindle Edition.

[158] Dennett, Daniel C. (2013-05-06). Intuition Pumps And Other Tools for Thinking (pp. 341-342). W. W. Norton & Company. Kindle Edition.

[159] Harris, Sam (2012-03-06). Free Will (p. 14). Simon & Schuster, Inc.. Kindle Edition.

[160] Critchley, Simon (2009-02-06). The Book of Dead Philosophers (Vintage) . Knopf Doubleday Publishing Group. Kindle Edition.

[161] Humphrey, Nicholas (2011-01-31). Soul Dust: The Magic of Consciousness (Kindle Locations 2227-2234). Princeton University Press. Kindle Edition.

[162] Tolle, Eckhart (2009-03-25). Stillness Speaks (Kindle Locations 217-219). New World Library. Kindle Edition.

[163] Hallinan, Joseph T. (2009-02-10). Why We Make Mistakes: How We Look Without Seeing, Forget Things in Seconds, and Are All Pretty Sure We Are Way Above Average (p. 202). Random House, Inc.. Kindle Edition.

[164] From Knocking on Heaven's Door, Katy Butler, Scribner, 2013, reported in the AARP Bulleting, October 2013, p. 38,

[165] Jonathan Haidt. The Happiness Hypothesis: Finding Modern Truth in Ancient Wisdom (Kindle Location 409). Kindle Edition.

[166] Humphrey, Nicholas (2011-01-31). Soul Dust: The Magic of Consciousness (Kindle Locations 2439-2442). Princeton University Press. Kindle Edition.

[167] http://www.mayoclinic.org/healthy-living/stress-management/in-depth/meditation/art-20045858

[168] Parfit, Derek.  Personal Identity.  The Philosophical Review, Vol. 80, No. 1 (Jan., 1971).  P.3.

[169] Jonathan Haidt. The Happiness Hypothesis: Finding Modern Truth in Ancient Wisdom (Kindle Locations 984-987). Kindle Edition.

[170] http://www.huffingtonpost.com/rabbi-manis-friedman/new-twist-old-story_b_2017349.html

[171] Tolle, Eckhart (2006-08-29). A New Earth (Oprah #61) (p. 9). Penguin Group. Kindle Edition.

[172] Jonathan Haidt. The Happiness Hypothesis: Finding Modern Truth in Ancient Wisdom (Kindle Locations 2643-2645). Kindle Edition.

[173] http://en.wikipedia.org/wiki/Peak_experience#cite_note-2

[174] Quine, Willard Van Orman.  Word & Object.  MIT Press.  1960. P3f.

[175] Parfit, Derek.  Reasons and Persons.  1986.  Oxford Paperbacks. P.281.

[176] Motesharrei, Safa, Rivas, Jorge, Kalnay, Eugenia.  Human and Nature Dynamics (HANDY):  Modeling Inequality and Use of Resources in the Collapse or Sustainability of Societies.  March 19, 2014, p.23.

90591801R00108

Made in the USA
Middletown, DE
25 September 2018